MEMOIRS
of the
American Mathematical Society
ISSN 0065-9266 (print)
ISSN 1947-6221 (online)

Number 1347

Instability, Index Theorem, and Exponential Trichotomy for Linear Hamiltonian PDEs

Zhiwu Lin
Chongchun Zeng

January 2022 • Volume 275 • Number 1347 (first of 6 numbers)

AMERICAN MATHEMATICAL SOCIETY

Library of Congress Cataloging-in-Publication Data

Cataloging-in-Publication Data has been applied for by the AMS.
See http://www.loc.gov/publish/cip/.
DOI: https://doi.org/10.1090/memo/1347

Memoirs of the American Mathematical Society

This journal is devoted entirely to research in pure and applied mathematics.

Subscription information. Beginning with the January 2010 issue, *Memoirs* is accessible from `www.ams.org/journals`. The 2022 subscription begins with volume 275 and consists of six mailings, each containing one or more numbers. Subscription prices for 2022 are as follows: for paper delivery, US$1085 list, US$868 institutional member; for electronic delivery, US$955 list, US$764 institutional member. Upon request, subscribers to paper delivery of this journal are also entitled to receive electronic delivery. If ordering the paper version, add US$22 for delivery within the United States; US$85 for outside the United States. Subscription renewals are subject to late fees. See `www.ams.org/help-faq` for more journal subscription information. Each number may be ordered separately; *please specify number* when ordering an individual number.

Back number information. For back issues see `www.ams.org/backvols`.

Subscriptions and orders should be addressed to the American Mathematical Society, P. O. Box 845904, Boston, MA 02284-5904 USA. *All orders must be accompanied by payment.* Other correspondence should be addressed to 201 Charles Street, Providence, RI 02904-2213 USA.

Memoirs of the American Mathematical Society (ISSN 0065-9266 (print); 1947-6221 (online)) is published bimonthly (each volume consisting usually of more than one number) by the American Mathematical Society at 201 Charles Street, Providence, RI 02904-2213 USA. Periodicals postage paid at Providence, RI. Postmaster: Send address changes to Memoirs, American Mathematical Society, 201 Charles Street, Providence, RI 02904-2213 USA.

This publication is indexed in *Mathematical Reviews®*, *Zentralblatt MATH*, *Science Citation Index®*, *Science Citation Index™-Expanded*, *ISI Alerting Services℠*, *SciSearch®*, *Research Alert®*, *CompuMath Citation Index®*, *Current Contents®/Physical, Chemical & Earth Sciences*.
This publication is archived in *Portico* and *CLOCKSS*.
Printed in the United States of America.

♾ The paper used in this book is acid-free and falls within the guidelines
established to ensure permanence and durability.
Visit the AMS home page at `https://www.ams.org/`

10 9 8 7 6 5 4 3 2 1 27 26 25 24 23 22

Contents

Abstract

Consider a general linear Hamiltonian system $\partial_t u = JLu$ in a Hilbert space X. We assume that $L : X \to X^*$ induces a bounded and symmetric bi-linear form $\langle L\cdot, \cdot\rangle$ on X, which has only finitely many negative dimensions $n^-(L)$. There is no restriction on the anti-self-dual operator $J : X^* \supset D(J) \to X$. We first obtain a structural decomposition of X into the direct sum of several closed subspaces so that L is blockwise diagonalized and JL is of upper triangular form, where the blocks are easier to handle. Based on this structure, we first prove the linear exponential trichotomy of e^{tJL}. In particular, e^{tJL} has at most algebraic growth in the finite co-dimensional center subspace. Next we prove an instability index theorem to relate $n^-(L)$ and the dimensions of generalized eigenspaces of eigenvalues of JL, some of which may be embedded in the continuous spectrum. This generalizes and refines previous results, where mostly J was assumed to have a bounded inverse. More explicit information for the indexes with pure imaginary eigenvalues are obtained as well. Moreover, when Hamiltonian perturbations are considered, we give a sharp condition for the structural instability regarding the generation of unstable spectrum from the imaginary axis. Finally, we discuss Hamiltonian PDEs including dispersive long wave models (BBM, KDV and good Boussinesq equations), 2D Euler equation for ideal fluids, and 2D nonlinear Schrödinger equations with nonzero conditions at infinity, where our general theory applies to yield stability or instability of some coherent states.

Received by the editor March 14, 2017, and, in revised form, January 28, 2018, and January 29, 2018.

Article electronically published on December 13, 2021.

DOI: https://doi.org/10.1090/memo/1347

2020 *Mathematics Subject Classification.* Primary 35B35, 37K45; Secondary 35P05, 47A10.

Key words and phrases. Linear Hamiltonian PDEs, stability, index formula, exponential trichotomy, structural decomposition, spectral mapping, Hamiltonian perturbations.

The first author is affiliated with the School of Mathematics, Georgia Institute of Technology, Atlanta, Georgia 30332. Email: zlin@math.gatech.edu.

The first author was supported in part by NSF-DMS 1411803 and NSF-DMS 1715201.

The second author is affiliated with the School of Mathematics, Georgia Institute of Technology, Atlanta, Georgia 30332. Email: chongchun.zeng@math.gatech.edu.

The second author was supported in part by NSF DMS-1362507.

CHAPTER 1

Introduction

We consider a general linear Hamiltonian system

$$\partial_t u = JLu, \ u \in X \tag{1.1}$$

in a real Hilbert space X. We assume that the operator $J : X^* \supset D(J) \to X$ satisfies $J^* = -J$ and $L : X \to X^*$ is bounded and satisfies $L^* = L$. This abstract equation is motivated by the linearization of a large class of Hamiltonian PDEs at equilibria or relative equilibria. Our first goal is to understand the structural and spectral properties of (1.1), its linear stability/instability, and the persistence of these properties under small perturbations in a general setting. Secondly, the general results on (1.1) will be applied to study the linearization at some coherent states of nonlinear Hamiltonian PDEs such as the 2-dim incompressible Euler equation, generalized Bullough-Dodd equation, Gross-Pitaevskii type equation, and some long wave models like KdV, BBM, and the good Boussinesq equations.

Our main assumption is that the quadratic form $\langle L\cdot, \cdot\rangle$ admits a decomposition $X = X_- \oplus \ker L \oplus X_+$, such that

$$\dim X_- = n^-(L) < \infty, \ \langle L\cdot, \cdot\rangle|_{X_-} < 0, \ \text{and} \ \langle L\cdot, \cdot\rangle|_{X_+} \geq \delta > 0.$$

An additional regularity assumption is required when $\dim \ker L = \infty$ (see **(H3)** in Section 2.1). We note that there is no additional restriction on the symplectic operator J, which can be unbounded, noninvertible, or even with infinite dimensional kernel.

* *Background: stability/instability and local dynamics near an equilibrium.* As our motivation for studying the linear system (1.1) is to understand the stability/instability of and the local dynamics near coherent states (steady states, traveling waves, standing waves etc.) of a nonlinear PDE, we first give a brief discussion of several standard notions of stability/instability and local dynamics. In a simple case of an ODE system

$$x_t = f(x), \quad x \in \mathbf{R}^n,$$

the local dynamics near an equilibrium x_0, without loss of generality assuming $x_0 = 0$, is very much related to the dynamics of its linearized equation

$$x_t = Ax, \quad A_{n\times n} = Df(0).$$

On the one hand, if A has an unstable eigenvalue λ ($\mathrm{Re}\,\lambda > 0$), then the above linearized equation has an exponential growing solution and is therefore linearly unstable. Here, linear stability means e^{tA} is uniformly bounded for all $t \geq 0$. While it is clearly linearly stable if $\mathrm{Re}\lambda < 0$ for all $\lambda \in \sigma(A)$, there might be linear solutions with polynomial growth if $\mathrm{Re}\lambda \leq 0$ for all $\lambda \in \sigma(A)$, which is often referred to as the spectrally stable case. Nonlinear instability immediately follows from spectral instability for ODEs. However, it is a much more subtle issue what properties in addition to the spectral (or even linear) stability would ensure nonlinear stability.

On the other hand, assume $\sigma_1 \subset \sigma(A)$ and $\mathrm{Re}\lambda < \alpha$ (or $\mathrm{Re}\lambda > \alpha$) for all $\lambda \in \sigma_1$. Let E_1 be the eigen-space of σ_1 which is invariant under e^{tA}, then we have the spectral mapping property

(SM) there exists $C > 0$ s.t. $|e^{tA}x| \leq Ce^{\alpha t}|x|$, $\forall x \in E_1$, $t \geq 0$ (or $t \leq 0$).

Suppose $\alpha_+ > \alpha_-$ and $\sigma(A) = \sigma_+ \cup \sigma_-$ with $\mathrm{Re}\lambda > \alpha_+$ for all $\lambda \in \sigma_+$ and $\mathrm{Re}\lambda < \alpha_-$ for all $\lambda \in \sigma_-$. Let $E_\pm$ be the eigen-spaces of $\sigma_\pm$, then the above spectral mapping property (SM) and $\alpha_+ > \alpha_-$ imply an *exponential dichotomy* of e^{tA}: in the decomposition $\mathbf{R}^n = E_+ \oplus E_-$ which is invariant under e^{tA}, the relative minimal exponential expanding rate of $e^{tA}|_{E_+}$ is greater than the maximal rate of $e^{tA}|_{E_-}$. For the nonlinear ODE system, the classical invariant manifold theory, based on the cornerstone of the exponential dichotomy, implies the existence of locally invariant (pseudo-)stable and unstable manifolds near 0. They often provide more detailed dynamic structures than the mere stability/instability and also help to organize the local dynamics.

It often happens that $f(x)$ and thus A depend on a small parameter ϵ, so one naturally desires to understand the dynamics of the perturbed systems for $0 < |\epsilon| << 1$ based on that of $\epsilon = 0$. A system is said to be structurally stable if its dynamics does not change qualitatively under any sufficiently small perturbation. For ODEs, it is well known that the local dynamics is structurally stable if A is hyperbolic, namely $\sigma(A) \cap i\mathbf{R} = \emptyset$.

The above ODE results may serve as *guidelines* in the study of local dynamics of PDEs near equilibria and relative equilibria while one has to keep in mind the following issues (among others):

- Sometimes it is highly non-trivial to analyze the spectra of linearized PDEs, particularly when the linear operator is not self-adjoint and has continuous spectrum.
- On the eigen-space E_1 of a spectral subset σ_1, the above spectral mapping type property (SM) may not hold for solutions of the linearized PDEs, due to the existence of continuous spectrum of the linearized operator (see e.g. [**66**]).
- Regularity issues in spatial variables can cause serious complications in proving nonlinear properties (stability/instability, local invariant manifolds, *etc.*) based on linear ones (spectral stability/instability, exponential dichotomy, *etc.*). The existing systematic results are mainly for semilinear PDEs.

* *Background: regarding Hamiltonian systems.* On a Hilbert space X, a Hamiltonian system takes the form

$$u_t = J\nabla H(u), \tag{1.2}$$

where the symplectic operator $J : X^* \to X$ satisfies $J^* = -J$ and $H : X \to \mathbf{R}$ is the Hamiltonian energy functional. In a more general setting, $J = J(u)$ may depend on u or (1.2) may be posed on a symplectic manifold M where $J(u) : T^*M \to TM$. In the classical setting, the symplectic structure $\omega \in T^*M \otimes T^*M$ is a 2-form given by

$$\omega(u)(U_1, U_2) = \langle J(u)^{-1}U_1, U_2\rangle, \ U_{1,2} \in T_uM,$$

which is required to be closed, namely $d\omega = 0$. It is standard that H and ω are invariant under the Hamiltonian flow associated with (1.2). Suppose u_* is a steady state of (1.2) (possibly in an appropriate reference frame, see examples in Chapter 11), then the linearized equation at u_* takes the form of (1.1) with $L = \nabla^2 H(u_*)$. In some cases, even though the nonlinear equation is not written in a straightforward Hamiltonian form, the linearization at an equilibrium u_* can still be put in the

Hamiltonian form (1.1), see Section 11.5 for the example of 2D Euler equation. It is standard for Hamiltonian ODEs and also proved for many Hamiltonian PDEs that the spectrum $\sigma(JL)$ is *symmetric* with respect to both real and imaginary axes. Therefore either (1.1) is spectrally unstable or its spectrum must lie on the imaginary axis. Even though the latter falls into the spectral stability category, it is often subtle to obtain properties of even the linear dynamics, such as linear stability and exponential dichotomy, based on the spectral properties, particular when there is continuous spectrum. Existing results in the literature often take advantage of the conservation of H or ω.

The structural stability is also more subtle even for linear Hamiltonian PDEs. On the one hand, the linearized operator JL associated with the linearization of Hamiltonian PDEs arising from physics and engineering usually has most of its spectrum lie on the imaginary axis. Therefore, the structural stability results based on the hyperbolicity of JL are hardly applicable. On the other hand, properties of Hamiltonian systems, such as the notions of Krein signatures and the conservation of H and ω, provide crucial additional tools. The structural stability of linear Hamiltonian PDEs addressed in this paper is mainly related to spectral properties and linear exponential dichotomy.

For Hamiltonian PDEs, there have been some works on local nonlinear dynamics based on properties of the linearized equations. For semilinear Hamiltonian PDEs $u_t = JH'(u)$ with nonlinear terms of subcritical growth, such as nonlinear Klein-Gordon equation, nonlinear Schrödinger equation, and Gross-Pitaevskii equation, local invariant manifolds can be constructed by combining ODE techniques with dispersive estimates (e. g. [**4**] [**38**] [**62**]). Such results for traveling wave solutions of the generalized KdV equation had also been obtained ([**37**]) with the help of smoothing estimates. The construction of invariant manifolds for quasilinear PDEs is more difficult, and was only done in very few cases (e. g. [**57**]). However, the passing from linear to nonlinear instability, which is a much weaker statement than the existence of invariant manifolds, had been done for many quasilinear PDEs (e.g. [**27**] [**31**] [**36**] [**50**] [**51**]). Several techniques were introduced to overcome the difficulties of loss of derivative of nonlinear terms and the growth due to the essential spectra of the linearized operators (see above references). The passing from spectral (or linear) stability to nonlinear stability is more subtle, particularly when $\langle Lu, u\rangle$ is not positive definite after the symmetry reduction. When such positivity holds, the nonlinear stability can usually be proved by using the Lyapunov functional, see e.g. [**29**] [**30**] for Hamiltonian PDEs. If such positivity fails, there is currently no general approach to study the nonlinear stability based on the linear one.

Our motivation of analyzing the linearized Hamiltonian system (1.1) in such a general form is to understand the stability/instability of and the local dynamics near a coherent state u_* of a nonlinear Hamiltonian PDE in the form of (1.2) with $L = \nabla^2 H(u_*)$. We first make some comments on the hypotheses.

On L, the assumption $n^-(L) < \infty$ is equivalent to that $H(u)$ has a finite Morse index at the critical point u_*. This assumption is automatically satisfied if u_* is constructed by minimizing $H(u)$ subject to finitely many constraints. In applications to continuum mechanics (fluids, plasmas etc.), the PDEs are often of a nonconomical Hamiltonian form $u_t = J(u)\nabla H(u)$, with a symplectic operator $J(u)$ depending on the solution u. In many cases, the linearization at an equilibrium

u_* can still be written in the Hamiltonian form (1.1) and the assumption $n^-(L) < \infty$ is satisfied (see Section 11.5 for the example of 2D Euler equation). The uniform positivity of L on X_+ could be relaxed to positivity by defining a new phase space (see Chapter 10).

In the existing literature on systems in the form of (1.1), $J^{-1} : X \to X^*$ is mostly assumed to be a bounded operator, which is not only for technical convenience but also natural in the sense that the symplectic 2-form ω is defined in terms of J^{-1}. However, it happens that J does not have a bounded inverse for many important Hamiltonian PDEs such as the KdV, BBM, the good Boussinesq equations, 2D Euler equation, *etc.*, see Chapter 11.

The goal of this paper regarding the general Hamiltonian PDE (1.1) is to study its spectral structures, linear dynamics, as well as certain structural stability properties under the assumption $n^-(L) < \infty$, but without any assumption on J in addition to $J^* = -J$. Our main general results include the symmetry of the spectrum $\sigma(JL)$, an index theorem relating certain spectral properties of JL to $n^-(L)$ which is useful for linear stability analysis, the linear exponential trichotomy of e^{tJL}, and the persistence of these properties for slightly perturbed Hamiltonian systems. These results are mostly achieved based on a structural decomposition of (1.1). In Chapter 11, several Hamiltonian PDEs are studied using these general results.

In the below, we briefly describe our main results and some key ideas in the proof. More details of the main theorems can be found in Chapter 2 and proofs in later chapters.

Structural decomposition. Most of the general theorems in this paper are based on careful decompositions of the phase space into closed subspaces through which L and JL take rather simple block forms. One of the most fundamental decomposition is given in Theorem 2.1. In this decomposition,

$$JL \longleftrightarrow \begin{pmatrix} 0 & A_{01} & A_{02} & A_{03} & A_{04} & 0 & 0 \\ 0 & A_1 & A_{12} & A_{13} & A_{14} & 0 & 0 \\ 0 & 0 & A_2 & 0 & A_{24} & 0 & 0 \\ 0 & 0 & 0 & A_3 & A_{34} & 0 & 0 \\ 0 & 0 & 0 & 0 & A_4 & 0 & 0 \\ 0 & 0 & 0 & 0 & 0 & A_5 & 0 \\ 0 & 0 & 0 & 0 & 0 & 0 & A_6 \end{pmatrix},$$

$$L \longleftrightarrow \begin{pmatrix} 0 & 0 & 0 & 0 & 0 & 0 & 0 \\ 0 & 0 & 0 & 0 & B_{14} & 0 & 0 \\ 0 & 0 & L_{X_2} & 0 & 0 & 0 & 0 \\ 0 & 0 & 0 & L_{X_3} & 0 & 0 & 0 \\ 0 & B_{14}^* & 0 & 0 & 0 & 0 & 0 \\ 0 & 0 & 0 & 0 & 0 & 0 & B_{56} \\ 0 & 0 & 0 & 0 & 0 & B_{56}^* & 0 \end{pmatrix},$$

where L takes an almost diagonal block form with $L_{X_3} \geq \delta$ for some $\delta > 0$ and JL takes a blockwise upper triangular form. Moreover all the blocks of JL are bounded operators except for A_3 which is anti-self-adjoint with respect to the equivalent inner product $\langle L_{X_3}\cdot, \cdot\rangle$ on X_3. In particular, all other diagonal blocks are matrices and therefore have only eigenvalues of finite multiplicity. The upper triangular form of JL simplifies the spectral analysis on JL tremendously and plays a fundamental role in the proof of the exponential trichotomy of e^{tJL}, the index formula, and the structural stability/instability of (1.1).

We briefly sketch some ideas in the construction of the decomposition here under the assumption $\ker L = \{0\}$, from which the decomposition in the general case follows. First, we observe that JL is anti-self-adjoint in the indefinite inner product $\langle L\cdot, \cdot\rangle$. Thus, by a Pontryagin type invariant subspace Theorem for symplectic operators in an indefinite inner product space, there exists an invariant (under JL) subspace $W \subset X$, satisfying that $L|_W \leq 0$ and $\dim W = n^-(L)$.

It would be highly *desirable* to extend W to a finite dimensional invariant subspace $\tilde{W}$ such that $L|_{\tilde{W}}$ is non-degenerate. This would yield the *invariant* decomposition $X = \tilde{W} \oplus \tilde{W}^{\perp L}$, where $\tilde{W}^{\perp L}$ is the orthogonal complement of $\tilde{W}$ with respect to $\langle L\cdot, \cdot\rangle$ and $L|_{\tilde{W}^{\perp L}} > 0$. Since $JL|_{\tilde{W}^{\perp L}}$ is anti-self-adjoint in the equivalent inner product $\langle L\cdot, \cdot\rangle$ and $\tilde{W}$ is finite dimensional, this immediately gives the decomposition we want.

However, such an invariant decomposition $X = \tilde{W} \oplus \tilde{W}^{\perp L}$ is in general impossible since it would imply that L is non-degenerate on the subspace of generalized eigenvectors of any purely imaginary eigenvalue of JL (Lemma 4.1), while the counterexample in Section 8.4 shows that L can be degenerate on such subspaces of embedded eigenvalues in the continuous spectra. Our proof is by a careful decomposition of the invariant spaces W, $W^{\perp L}$ and their complements.

Exponential trichotomy. Our second result is the exponential trichotomy of e^{tJL} in X and more regular spaces (Theorem 2.2). More precisely, we decompose $X = E^u \oplus E^c \oplus E^s$, such that: $E^{u,c,s}$ are invariant under e^{tJL},

$$\dim E^u = \dim E^s \leq n^-(L), \ E^c = (E^u \oplus E^s)^{\perp L},$$

and $A_5 = e^{tJL}|_{E^u}$ $(A_6 = e^{tJL}|_{E^s})$ has exponential decay when $t < 0$ $(t > 0)$ and $e^{tJL}|_{E^c}$ has possible polynomial growth for all t with the optimal algebraic rate explicitly given. Roughly speaking, the unstable (stable) spaces E^u (E^s) are subspaces of generalized eigenvectors of the unstable (stable) eigenvalues of JL and the center space E^c corresponds to the spectra in the imaginary axis.

Such exponential trichotomy is an important step to prove nonlinear instability, and furthermore to construct local invariant (stable, unstable, center) manifolds which are crucial for a complete understanding of the local dynamics, see, for example, [**4, 16, 17**]. Such exponential trichotomy or dichotomy might be tricky to get due to the spectral mapping issue, that is, generally $\sigma\left(e^{tJL}\right) \subsetneq e^{t\sigma(JL)}$. So even if the spectra of JL is understood, it is still a subtle issue to prove the estimates for e^{tJL}. In the literature, the exponential dichotomy is usually obtained either by resolvent estimates (e.g. [**26**]) or compact perturbations of simpler semigroups ([**72**] [**67**]). The proofs were often technical (particularly for resolvent estimates) and only worked for specific classes of problems. Our result gives the exponential trichotomy for general Hamiltonian PDEs (1.1) with $n^-(L) < \infty$. Moreover, the growth rates (particularly on the center space) obtained are sharp. In particular, our sharp polynomial growth rate estimate on the center space implies a stronger result than the usual spectral mapping statement. Our proof of the exponential trichotomy which is very different from traditional methods, is based on the upper triangular form of JL in the decomposition given in Theorem 2.1. It can be seen that the Hamiltonian structure of (1.1) plays an important role in the proof.

Index theorems. Our third result is an index formula to relate the counting of dimensions of some eigenspaces of JL to $n^-(L)$. Denote the sum of algebraic multiplicities of all positive eigenvalues of JL by k_r and the sum of algebraic multiplicities of eigenvalues of JL in the first quadrant by k_c. Let $k_i^{\leq 0}$ be the total number of *nonpositive* dimensions $n^{\leq 0}(L|_{E_{i\mu}})$ of the quadratic form $\langle L\cdot,\cdot\rangle$ restricted to the subspaces $E_{i\mu}$ of generalized eigenvectors of all purely imaginary eigenvalues $i\mu \in \sigma(JL) \cap i\mathbf{R}$ of JL with positive imaginary parts, and $k_0^{\leq 0}$ be the number of nonpositive dimensions of $\langle L\cdot,\cdot\rangle$ restricted to the generalized kernel of JL modulo $\ker L$. We note that, when all purely imaginary eigenvalues are semi-simple and $\langle L\cdot,\cdot\rangle$ restricted to these kernels is non-degenerate, $k_i^{\leq 0}$ is equal to k_i^- which represents the number of purely imaginary eigenvalues (with positive imaginary parts) of negative Krein signature. The situation is more complicated if the eigenvalue is not semi-simple or even embedded into the continuous spectra. In the general case, we have

$$k_r + 2k_c + 2k_i^{\leq 0} + k_0^{\leq 0} = n^-(L)\,. \tag{1.3}$$

Two immediate corollaries of (1.3) are: $n^-(L) = k_0^{\leq 0}$ implies spectral stability and the oddness of $n^-(L) - k_0^{\leq 0}$ implies linear instability. Since by (1.3) all the negative directions of $\langle L\cdot,\cdot\rangle$ are associated to eigenvalues of JL, conceptually the continuous spectrum of JL is only associated to positive directions of $\langle L\cdot,\cdot\rangle$.

There have been lots of work on similar index formulae under various settings in the literature. In the finite dimensional case where L and JL are matrices, such index formula readily follows from arguments in a paper of Mackay [**59**], although was not written explicitly there. In the past decade, there have been lots of work trying to extend it to the infinite dimensional case. In most of these papers, J is assumed to have a bounded inverse ([**18**] [**21**] [**44**] [**47**]), or $J|_{(\ker J)^\perp}$ has a bounded inverse, as in the cases of periodic waves of dispersive PDEs ([**11**] [**13**] [**32**] [**43**]). Recently, in [**46**] [**65**], the index formulae were studied for KDV type equations in the whole line for which $J = \partial_x$ does not have bounded inverse. Our result (1.3) gives a generalization of these results since we allow J to be an arbitrary anti-self-dual operator. In particular, $J|_{(\ker J)^\perp}$ does not need to have a bounded inverse. This is important for applications to continuum mechanics (e.g. fluids and plasmas) where J usually has an infinite dimensional kernel with 0 in the essential spectrum of J in some appropriate sense (see Section 11.5 for the example of 2D Euler equation).

We should also point out some differences of (1.3) with previous index formulae even in the case with bounded J^{-1}. In previous works on index formula, it is assumed that $\langle L\cdot,\cdot\rangle$ is non-degenerate on $(JL)^{-1}(\ker L)/\ker L$. Under this assumption, the generalized kernel of JL only have Jordan blocks of length 2 and $k_0^{\leq 0} = n^-\left(L|_{(JL)^{-1}(\ker L)/\ker L}\right)$ (see Propositions 2.7 and 2.8). In (1.3), we do not impose such non-degeneracy assumption on $L|_{(JL)^{-1}(\ker L)/\ker L}$ and thus the possible structures may be much richer. In the counting of (1.3), we use $k_i^{\leq 0}, k_0^{\leq 0}$, which are the total dimensions of *non-positive* directions of L restricted on the subspaces $E_{i\mu}$ of generalized eigenvectors of purely imaginary eigenvalues $i\mu$ or zero eigenvalue (modulo $\ker L$). Since $\langle L\cdot,\cdot\rangle$ might be degenerate on such subspace $E_{i\mu}$ of an embedded eigenvalue (see example in Section 8.4), they can not be replaced by k_i^-, k_0^- (i.e. the dimensions of negative directions of L) as used in the index formula of some papers (e.g. [**44**]). However, in Proposition 2.3, we show that if a purely imaginary spectral point $i\mu$ is isolated, then L is non-degenerate on its generalized eigenspace $E_{i\mu}$ which consists of generalized eigenvectors only. In this case, we also get an explicit formula (2.16) for $n^{-}\left(L|_{E_{i\mu}}\right)$ by its Jordan canonical form, which is independent of the choice of the basis realizing the canonical form. This formula suggests that even for embedded eigenvalues which might be of infinite multiplicity, the number and length of nontrivial Jordan chains are bounded in terms of $n^{-}(L)$.

Moreover, even for the case where $\langle L\cdot,\cdot\rangle$ is degenerate on $E_{i\mu}$, we give a block decomposition of JL and L on $E_{i\mu}$ (Proposition 2.2). In this decomposition, L is blockwise diagonal and JL takes an upper triangular form with three diagonal blocks corresponding to the degenerate part of L, the simple eigenspaces and the Jordan blocks of $i\mu$ of JL. Furthermore,we construct a special basis for each Jordan block such that the corresponding L is in an anti-diagonal form (2.15). The above decomposition of $E_{i\mu}$ yields formula (2.16) for the case where $\langle L\cdot,\cdot\rangle|_{E_{i\mu}}$ is non-degenerate and also plays an important role on the constructive proof of Pontryagin type invariant subspace Theorem 5.1 and the proof of structural instability Theorem 2.6. To our knowledge, the formula (2.16) and the decomposition in Proposition 2.2 are new even for the finite dimensional case.

We also note that for an eigenvalue λ with $\operatorname{Re}\lambda \neq 0$, $\langle L\cdot,\cdot\rangle|_{E_\lambda} = 0$ and by Corollary 6.1 $\langle L\cdot,\cdot\rangle|_{E_\lambda\oplus E_{-\bar\lambda}}$ is non-degenerate with

$$n^{-}\left(L|_{E_\lambda\oplus E_{-\bar\lambda}}\right) = \dim E_\lambda. \tag{1.4}$$

Therefore, we get the matrix form

$$\langle L\cdot,\cdot\rangle|_{E_\lambda\oplus E_{-\bar\lambda}} \longleftrightarrow \begin{pmatrix} 0 & A \\ A^* & 0 \end{pmatrix},$$

where A is a nonsingular $n\times n$ matrix with $n = \dim E_\lambda$.

Now we discuss some ideas in our proof of index formula and the decomposition in Proposition 2.2 after we briefly review previous approaches for the index formulae. Like in the literature ([**44**] [**21**]), the index formula was usually proved by reducing the eigenvalue problem $JLu = \lambda u$ to a generalized eigenvalue problem $(R - zS)v = 0$ (so called linear operator pencil), where $z = -\lambda^2$ and R, S are self-adjoint operators with $\ker S = \{0\}$. To get such reduction it is required that J has a bounded inverse and L is non-degenerate on $(JL)^{-1}(\ker L)/\ker L$. Notice that the operator $S^{-1}R$ is self-adjoint in the indefinite inner product $\langle S\cdot,\cdot\rangle$. So by the Pontryagin invariant subspace theorem ([**28**] [**18**] [**48**] [**64**]) for self-adjoint operators, there is an $n^{-}(S)$-dimensional invariant (under $S^{-1}R$) subspace W such that $\langle S\cdot,\cdot\rangle|_W \leq 0$, where

$$n^{-}(S) = n^{-}(L) - n\left(L|_{(JL)^{-1}(\ker L)/\ker L}\right).$$

Going back to the original problem $JLu = \lambda u$, an index formula can be obtained by counting the negative dimensions of L on the eigenspaces for real, complex and pure imaginary eigenvalues. However, it should be pointed out that the counting in some papers used the formula (1.4), for which the required non-degeneracy of $L|_{E_\lambda\oplus E_{-\bar\lambda}}$ seemed to be assumed but not proved.

In [**32**] and later also in [**13**] [**11**] [**43**], the index formula was proved without reference to the Pontryagin invariant subspace theorem. In these papers, some conditions on J and L were imposed to ensure that the generalized eigenvectors of JL form a complete basis of X. Then the index formula follows by the arguments as in the finite dimensional case ([**59**]). Such requirement of a complete basis is very strong and mostly true only in some cases where the eigenvalues of JL are all discrete.

Our proof of the index formula (1.3) is based on the decomposition in Theorem 2.1, where we used the Pontryagin invariant subspace theorem for the anti-self-adjoint operator JL in the indefinite inner product $\langle L\cdot,\cdot\rangle$. The proof of the detailed decompositions of JL and L on $E_{i\mu}$ given in Proposition 2.2, particularly the construction of the special basis realizing the Jordan canonical form, is carried out in two steps. First, in the finite dimensional case, we construct a special basis of the eigenspace $E_{i\mu}$ of JL to skew-diagonalize L on the Jordan blocks by using an induction argument on the length of Jordan chains. Second, for the infinite dimensional case, we decompose $E_{i\mu}$ into subspaces corresponding to degenerate eigenspaces, simple non-degenerate eigenspaces and Jordan blocks. Since the Jordan block part is finite dimensional, the special basis is constructed as in the finite dimensional case.

Hamiltonian perturbations. Our fourth main result is about the persistence of exponential trichotomy and a sharp condition for the structural stability of linear Hamiltonian systems under small Hamiltonian perturbations. Consider a perturbed Hamiltonian system $u_t = J_\# L_\# u$ where $J_\#, L_\#$ are small perturbations of J, L in the sense of (2.24). This happens when the symplectic structure or the Hamiltonian of the system depends on some parameters.

First, we show that the exponential trichotomy of e^{tJL} persists under small perturbations. More precisely, we show in Theorem 2.4 that there exists a decomposition $X = E^u_\# \oplus E^s_\# \oplus E^c_\#$, satisfying that: $E^{u,s,c}_\#$ are invariant under $e^{tJ_\# L_\#}$ and are obtained as small perturbations of $E^{u,s,c}$ in the sense that $E^{u,s,c}_\# = \operatorname{graph}(S^{u,s,c}_\#)$ where

$$S^u_\# : E^u \to E^s \oplus E^c, \quad S^s_\# : E^s \to E^u \oplus E^c, \quad S^c_\# : E^c \to E^s \oplus E^u, \quad |S^{u,s,c}_\#| \le C\epsilon,$$

and ϵ is roughly the size of perturbations $L_\# - L$ and $J_\# - J$ (see (2.24)). Moreover, $e^{tJ_\# L_\#}$ has exponential decay on $E^u_\#$ and $E^s_\#$ in negative and positive times respectively with at most $O(\epsilon)$ loss of decay rates compared with $e^{tJL}|_{E^{u,s}}$; on $E^c_\#$, $e^{tJ_\# L_\#}$ has at most small exponential growth at the rate $O(\epsilon)$. We note that $J_\# L_\#|_{E^c_\#}$ might contain eigenvalues with small real parts which are perturbed from the spectra of JL in the imaginary axis and thus the small exponential growth on $e^{tJ_\# L_\#}$ is the best one can get. In the perturbed decomposition $E^{u,s,c}_\#$, we obtain the uniform control of the growth rate and the bounds in semigroup estimates for $e^{tJ_\# L_\#}$ on $E^{u,s,c}_\#$. Such uniform estimates of the exponential trichotomy (or dichotomy) are important for many applications of nonlinear perturbation problems, such as the modulational instability of dispersive models (see Lemma 11.2).

We briefly discuss some ideas in the proof of Theorem 2.4. The spaces $E^{u,s}_\#$ are constructed as the ranges of the projection operators $\tilde{P}^{u,s}_\#$ by the Riesz projections associated with the operator $J_\# L_\#$ in a contour enclosing $\sigma(JL|_{E^{u,s}})$ and $E^c_\# = \left(E^{u,s}_\#\right)^{\perp L_\#}$. The smallness assumption (2.24) is used in the resolvent estimates to show that $E^{u,s,c}_\#$ are indeed $O(\epsilon)$ perturbations of $E^{u,s,c}$. It is actually not so straightforward to prove the small exponential growth of $e^{tJ_\# L_\#}$ on $E^c_\#$ since the perturbation term $J(L_\# - L)$ may be unbounded. We again use the decomposition Theorem 2.1, where in the decomposition for JL, only one block is infinite dimensional, with good structure, and others blocks are all bounded.

In Theorems 2.5 and 2.6, we prove that a pure imaginary eigenvalue $i\mu \neq 0$ of JL is structurally stable, in the sense that the spectra of $J_{\#}L_{\#}$ near $i\mu$ stay in the imaginary axis, *if and only if* either $L|_{E_{i\mu}} > 0$ or $i\mu$ is isolated and $L|_{E_{i\mu}} < 0$. In particular, when $\langle L\cdot,\cdot\rangle$ is indefinite on $E_{i\mu}$ or $i\mu$ is an embedded eigenvalue and $\langle Lu,u\rangle \leq 0$ for some $0 \neq u \in E_{i\mu}$, there exist perturbed operators $JL_{\#}$ with unstable eigenvalues near $i\mu$ and $|L_{\#} - L|$ being arbitrarily small. The structural stability of finite dimensional Hamiltonian systems had been well studied in the literature (see **[24]** **[59]** and references therein). It was known that (see e.g. **[59]**) a purely imaginary eigenvalue $i\mu \neq 0$ is structurally stable if and only if L is definite on $E_{i\mu}$. As a consequence, for a family of Hamiltonian systems, the equilibrium can lose spectral stability only by the collision of purely imaginary eigenvalues of opposite Krein signatures (i.e. sign of $\langle L\cdot,\cdot\rangle$) . For Hamiltonian PDEs, the situation is more subtle due to the possible embedded eigenvalues in the continuous spectrum. In **[28]**, the linearized equation at excited states of a nonlinear Schrödinger equation was studied and the structural instability was shown for an embedded simple eigenvalue with negative signature. A similar result was also obtained in **[21]** for semi-simple embedded eigenvalues. The assumptions in Theorems 2.5 and 2.6 are much more general and they give a sharp condition for the structural stability of nonzero pure imaginary eigenvalues of general Hamiltonian operator JL. In particular, in Theorem 2.6, structural instability is proved even for the case when the embedded eigenvalue is degenerate, which was not included in **[28]** or **[21]** for linearized Schrödinger equations.

In the below, we discuss some ideas in the proof of Theorems 2.5 and 2.6. In the finite dimensional case, the structural stability of an eigenvalue $i\mu$ of JL with a definite energy quadratic form $L|_{E_{i\mu}}$ can be readily seen from an argument based on Lyapunov functions. The above intuition can be used to show structural stability in Theorem 2.5 for isolated eigenvalues with definite energy quadratic forms. The proof is more subtle for embedded eigenvalues with positive energy quadratic forms. We argue via contradiction by showing that if there is a sequence of unstable eigenvalues perturbed from $i\mu$, then this leads to a non-positive direction of $L|_{E_{i\mu}}$. In this proof, the decomposition Theorem 2.1 again plays an important role. The proof of structural instability Theorem 2.6 is divided into several cases. When $L|_{E_{i\mu}}$ is non-degenerate and indefinite, it can be reduced to the finite dimensional case for which we can construct a perturbed matrix to have unstable eigenvalues. In particular, in the case when $E_{i\mu}$ contains a Jordan chain on which L is non-degenerate, we use the special basis in Proposition 2.2 to construct a perturbed matrix with unstable eigenvalues.

The proof is more subtle for an embedded eigenvalue $i\mu$ with non-positive and possibly degenerate $\langle L\cdot,\cdot\rangle\,|_{E_{i\mu}}$. First, we construct a perturbed Hamiltonian system $J\tilde{L}_{\#}$ near JL such that $i\mu$ is an isolated eigenvalue of $J\tilde{L}_{\#}$ and there is a positive direction of $\tilde{L}_{\#}|_{E_{i\mu}(J\tilde{L}_{\#})}$. In this construction, we use the decomposition Theorem 2.1 once again along with spectral integrals. Then by Proposition 2.3, $\tilde{L}_{\#}|_{E_{i\mu}(J\tilde{L}_{\#})}$ is non-degenerate and is indefinite by our construction. Thus it is reduced to the previously studied cases. In a rough sense, the structural instability is induced by the resonance between the embedded eigenvalue (with $\langle L\cdot,\cdot\rangle$ non-positive in the directions of some generalized eigenvectors) and the pure continuous spectra whose spectral space has only positive directions due to the index formula (1.3).

In some applications (see e.g. Section 11.6), it is not easy to get the *uniform* positivity for $L|_{X^+}$ (i.e. assumption (**H2.b**)) in an obvious space X and only the positivity $L|_{X^+}$ is available. In Theorem 2.7, we show that under some additional assumptions ((**B1**)-(**B5**) in Section 2.6), one can construct a new phase space Y such that X is densely embedded into Y; the extension L_Y of L satisfies the uniform positivity in $\|\cdot\|_Y$; $J_Y : D(J) \cap Y^* \to Y$ is the restriction of J, and (J_Y, L_Y, Y) satisfy the main assumptions (**H1-3**). Then we can apply the theorems to (J_Y, L_Y, Y).

Hamiltonian PDE models. In Chapter 11 (see also Section 2.7 for a summary), we study the stabilities and related issues of various concrete Hamiltonian PDEs based on our above general theory, including: stability of solitary and periodic traveling waves of long wave models of BBM, KDV, and good Boussinesq types; the eigenvalue problem of the form $Lu = \lambda u'$ arising from the stability of solitary waves of generalized Bullough–Dodd equation; modulational instability of periodic traveling waves; stability of steady flows of 2D Euler equations; traveling waves of 2D nonlinear Schrödinger equations with nonzero condition at infinity.

This paper is organized as follows. In Chapter 2, we give the precise set-up and list the main general results more precisely with some comments, where the readers are directed to the corresponding subsequent sections for detailed proofs. *For some readers, who would like to see the general results but do not desire to get into the technical details of the proofs, it is possibly sufficient to read Sections 2.1–2.6 only.* The stability analysis of various Hamiltonian PDEs are outlined in Section 2.7. The proofs of the main general results are given in Chapters 3 to 10. Chapter 3 studies some basic properties of linear Hamiltonian systems. Chapter 4 is about the finite dimensional Hamiltonian systems. In particular, the special basis in Proposition 2.2 is constructed. Chapter 5 is about the Pontryagin type invariant subspace Theorem for anti-self-adjoint operators in an indefinite inner product space. Two proofs are given. One is by the fixed point argument as found in the literature ([**18**] [**25**] [**48**]), which provides the existence of an invariant Pontryagin subspace abstractly. The second one in separable Hilbert spaces is via Galerkin approximation which also yields an *explicit* construction of a maximally non-positive invariant subspace. Chapter 6 is to prove decomposition Theorem 2.1 which plays a crucial role in the proof of most of the main results. Chapter 7 contains the proof of the exponential trichotomy of e^{tJL}. In Chapter 8, the index theorem is proved. Besides, the structures of the generalized eigenspaces are studied and more explicit formula for the indexes $k_i^{\leq 0}, k_0^{\leq 0}$, etc. are proved. The non-degeneracy of $L|_{E_{i\mu}}$ for any isolated spectral point $i\mu$ is also proved there. In Chapter 9, we prove the persistence of the exponential trichotomy and the structural stability/instability Theorems. In Chapter 10, we prove that the uniform positivity assumption (**H2.b**) can be relaxed under some assumptions. We study the stability and related issues of various Hamiltonian PDEs in Chapter 11. In the Appendix, we prove some functional analysis facts used throughout the paper, including some basic decompositions of the phase space, the well-posedness of the linear Hamiltonian system, and the standard complexification procedure.

CHAPTER 2

Main Results

In this chapter, we give details of the main results described in the introduction. The detailed proofs are left for later sections.

A remark on notations: Throughout the paper, given a densely defined linear operator T from a Banach space X to a Banach space Y we will always use T^* to denote its dual operator from a subspace of Y^* to X^*. It would *never* mean the adjoint operator even if $X = Y$ is a Hilbert space. Given a Hilbert space X and a linear operator $L : X \to X^*$, since $L^* : (X^*)^* = X \to X^*$, it is legitimate to compare whether $L = L^*$.

2.1. Set-up

Consider a linear Hamiltonian system

$$\partial_t u = JLu, \quad u \in X \tag{2.1}$$

where X is a real Hilbert space. Let $(\cdot, \cdot)$ denote the inner product on X and $\langle \cdot, \cdot \rangle$ the dual bracket between X^* and X. We make the following assumptions:

- **(H1)** $J : X^* \supset D(J) \to X$ is anti-self-dual, in the sense $J^* = -J$.
- **(H2)** The operator $L : X \to X^*$ is bounded and symmetric (i.e. $L^* = L$) such that $\langle Lu, v \rangle$ is a bounded symmetric bilinear form on X. Moreover, there exists a decomposition of X into the direct sum of three closed subspaces

$$X = X_- \oplus \ker L \oplus X_+, \quad n^-(L) \triangleq \dim X_- < \infty$$

satisfying

- **(H2.a)** $\langle Lu, u \rangle < 0$ for all $u \in X_- \backslash \{0\}$;
- **(H2.b)** there exists $\delta > 0$ such that

$$\langle Lu, u \rangle \geq \delta \|u\|^2 \ , \text{ for any } u \in X_+.$$

- **(H3)** The above $X_\pm$ satisfy

$$\ker i^*_{X_+ \oplus X_-} = \{f \in X^* \mid \langle f, u \rangle = 0, \forall u \in X_- \oplus X_+\} \subset D(J)$$

where $i^*_{X_+ \oplus X_-} : X^* \to (X_+ \oplus X_-)^*$ is the dual operator of the embedding $i_{X_+ \oplus X_-}$.

REMARK 2.1. *If in addition we assume*

$$\ker i^*_{(\ker L)^\perp} = \{f \in X^* \mid \langle f, u \rangle = 0, \forall u \in (\ker L)^\perp\} \subset D(J) \tag{2.2}$$

where

$$(\ker L)^\perp = \{u \in X \mid (u, v) = 0, \ \forall v \in \ker L\}, \tag{2.3}$$

it is possible to choose $X_\pm \subset (\ker L)^\perp$. See Lemma A.4 *and Remark* A.4 *in the Appendix.*

Regarding the operator L, what often matters more is its associated symmetric quadratic form $\langle Lu, v\rangle$, $u, v \in X$, (or the Hermitian symmetric form after the complexification). We say a bounded symmetric quadratic form $B(u, v)$ is *non-degenerate* if

$$\inf_{v\neq 0} \sup_{u\neq 0} \frac{|B(u,v)|}{\|u\|\|v\|} > 0, \tag{2.4}$$

or equivalently, $v \to f = B(\cdot, v) \in X^*$ defines an isomorphism from X to X^* (or a complex conjugate (sometimes called anti-linear) isomorphism – satisfying $av \to \bar{a} f$ for any $a \in \mathbf{C}$ – after the complexification). Under assumptions (**H1-3**), $\langle Lu, v\rangle$ is non-degenerate if and only if $\ker L = \{0\}$ (see Lemma A.2).

REMARK 2.2. *It is worth pointing out that* $n^-(L) = \dim X_-$ *is actually the maximal dimension of subspaces where* $\langle L\cdot, \cdot\rangle < 0$, *see Lemma* A.1. *Thus* $n^-(L)$ *is the Morse index of* L.

By Riesz Representation Theorem, there exists a unique bounded symmetric linear operator $\mathbb{L} : X \to X$ *such that* $(\mathbb{L}u, v) = \langle Lu, v\rangle$. *Let* Π_λ, $\lambda \in \mathbf{R}$, *denote the orthogonal spectral projection operator from* X *to the closed subspace corresponding to the spectral subset* $\sigma(\mathbb{L}) \cap (-\infty, \lambda]$. *From the standard spectral theory of self-adjoint operators, assumption* ***(H2)*** *is equivalent to that there exists* $\delta' > 0$ *such that*
i.) $\sigma(\mathbb{L}) \cap [-\delta', \delta'] \subset \{0\}$, *which is equivalent to the closeness of* $R(L)$, *and*
ii.) $\dim(\Pi_{-\delta'} X) < \infty$.
The subspaces

$$X_- = \Pi_{-\frac{\delta'}{2}} X \qquad X_+ = (I - \Pi_{\frac{\delta'}{2}})X,$$

along with $\ker L$ *lead to a decomposition of* X *orthogonal with respect to both* $(\cdot, \cdot)$ *and* $\langle L\cdot, \cdot\rangle$, *satisfying* ***(H2)***.

REMARK 2.3. *We would like to point out that* ***(H3)*** *is automatically satisfied if* $\dim \ker L < \infty$. *In fact in this case,*

$$\dim \ker i^*_{X_+ \oplus X_-} = \dim\{f \in X^* \mid \langle f, u\rangle = 0, \ \forall u \in X_- \oplus X_+\} = \dim \ker L < \infty.$$

Let $\{f_1, \ldots, f_k\}$ *be a basis of* $\ker i^*_{X_+\oplus X_-}$. *As* $D(J)$ *is dense in* X^*, *one may take* $g_j \in D(J)$ *sufficiently close to* f_j, $j = 1, \ldots, k$. *Let*

$$X_1 = \{u \in X \mid \langle g_j, u\rangle = 0, \ \forall j = 1, \ldots, k\}.$$

Since X_1 *is close to* $X_+ \oplus X_-$, *it is easy to show that there exist closed subspaces* $X_{1\pm} \subset X_1$ *satisfying* ***(H2)*** *and* $X_1 = X_{1+} \oplus X_{1-}$.

In fact, if we had treated L *and* J *as operators from* X *to* X *through the Riesz Representation Theorem and* $X_\pm$ *happen to be given as in Remark* 2.2 *then* ***(H3)*** *would take the form* $\ker L \subset D(J)$.

Assumption ***(H3)*** *does ensure that* JL *is densely defined, see Lemma* A.5.

REMARK 2.4. *Assumption* ***(H2.b)*** *requires that the quadratic form* $\langle Lu, u\rangle$ *has a uniform positive lower bound on* X_+. *This corresponds to that* 0 *is an isolated eigenvalue of* $\mathbb{L}$ *defined in Remark* 2.2, *which also implies that* $R(L)$ *is closed and* $R(L) = \{\gamma \in X^* \mid \langle\gamma, u\rangle = 0, \forall u \in \ker L\}$.

For some PDE systems, ***(H2.b)*** *may not hold or be hard to verify, see, e.g. Section* 11.6. *In Section* 2.6, *we consider a framework where assumption* ***(H2.b)*** *for the uniform positivity of* $L|_{X_+}$ *is weakened to the positivity of* $L|_{X_+}$, *if some* *additional and more detailed structures are present. In that situation, we construct a new phase space $Y \supset X$ and extend the operators L and J to Y accordingly so that (**H1-3**) are satisfied.*

2.2. Structural decomposition

Our first main result is to construct a decomposition of the phase space X which helps understanding both structures of JL and L simultaneously.

THEOREM 2.1. *Assume (**H1-H3**). There exist closed subspaces X_j, $j=1,\dots,6$, and $X_0 = \ker L$ such that*

(1) $X = \oplus_{j=0}^{6} X_j$, $X_j \subset \cap_{k=1}^{\infty} D\big((JL)^k\big)$, $j \neq 3$, *and*

$$\dim X_1 = \dim X_4,\ \dim X_5 = \dim X_6,\ \dim X_1 + \dim X_2 + \dim X_5 = n^-(L);$$

(2) *JL and L take the following forms in this decomposition*

$$JL \longleftrightarrow \begin{pmatrix} 0 & A_{01} & A_{02} & A_{03} & A_{04} & 0 & 0 \\ 0 & A_1 & A_{12} & A_{13} & A_{14} & 0 & 0 \\ 0 & 0 & A_2 & 0 & A_{24} & 0 & 0 \\ 0 & 0 & 0 & A_3 & A_{34} & 0 & 0 \\ 0 & 0 & 0 & 0 & A_4 & 0 & 0 \\ 0 & 0 & 0 & 0 & 0 & A_5 & 0 \\ 0 & 0 & 0 & 0 & 0 & 0 & A_6 \end{pmatrix}, \tag{2.5}$$

$$L \longleftrightarrow \begin{pmatrix} 0 & 0 & 0 & 0 & 0 & 0 & 0 \\ 0 & 0 & 0 & 0 & B_{14} & 0 & 0 \\ 0 & 0 & L_{X_2} & 0 & 0 & 0 & 0 \\ 0 & 0 & 0 & L_{X_3} & 0 & 0 & 0 \\ 0 & B_{14}^* & 0 & 0 & 0 & 0 & 0 \\ 0 & 0 & 0 & 0 & 0 & 0 & B_{56} \\ 0 & 0 & 0 & 0 & 0 & B_{56}^* & 0 \end{pmatrix}. \tag{2.6}$$

(3) *$B_{14} : X_4 \to X_1^*$ and $B_{56} : X_6 \to X_5^*$ are isomorphisms and there exists $\delta > 0$ satisfying $\mp\langle L_{X_{2,3}} u, u\rangle \geq \delta \|u\|^2$, for all $u \in X_{2,3}$;*
(4) *all blocks of JL are bounded operators except A_3, where A_{03} and A_{13} are understood as their natural extensions defined on X_3;*
(5) *$A_{2,3}$ are anti-self-adjoint with respect to the equivalent inner product $\mp\langle L_{X_{2,3}}\cdot,\cdot\rangle$ on $X_{2,3}$;*
(6) *the spectra $\sigma(A_j) \subset i\mathbf{R}$, $j = 1,2,3,4$, $\pm \operatorname{Re}\lambda > 0$ for all $\lambda \in \sigma(A_{5,6})$, and $\sigma(A_5) = -\sigma(A_6)$;*
(7) *$n^-(L|_{X_5\oplus X_6}) = \dim X_5$ and $n^-(L|_{X_1\oplus X_4}) = \dim X_1$.*
(8) *$(u,v) = 0$ for all $u \in X_1 \oplus X_2 \oplus X_3 \oplus X_4$ and $v \in \ker L$.*

Through straightforward calculations, one may naturally rewrite the operator J and obtain additional relations among those blocks A_{jk} using $J^* = -J$.

COROLLARY 2.1. *Let P_j, $j = 0,\dots,6$ be the projections associated to the decomposition in Theorem 2.1 and $\tilde{X}_j^* = P_j^* X_j^* \subset X^*$. In the decomposition*

$X^* = \Sigma_{j=0}^6 \tilde{X}_j^*$, *J has the block form*

$$J \longleftrightarrow \begin{pmatrix} J_{00} & J_{01} & J_{02} & J_{03} & J_{04} & 0 & 0 \\ J_{10} & J_{11} & J_{12} & J_{13} & J_{14} & 0 & 0 \\ J_{20} & J_{21} & J_{22} & 0 & 0 & 0 & 0 \\ J_{30} & J_{31} & 0 & J_{33} & 0 & 0 & 0 \\ J_{40} & J_{41} & 0 & 0 & 0 & 0 & 0 \\ 0 & 0 & 0 & 0 & 0 & 0 & J_{56} \\ 0 & 0 & 0 & 0 & 0 & J_{65} & 0 \end{pmatrix}.$$

where the blocks, except J_{00}, *are given by*

$$\begin{aligned} &-J_{10}^* = J_{01} = A_{04}B_{14}^{-1}, \quad -J_{20}^* = J_{02} = A_{02}L_2^{-1} \\ &-J_{30}^* = J_{03} = A_{03}L_{X_3}^{-1}, \quad -J_{40}^* = J_{04} = A_{01}(B_{14}^*)^{-1} \\ &J_{11} = A_{14}B_{14}^{-1}, \quad J_{12} = A_{12}L_{X_2}^{-1}, \quad J_{13} = A_{13}L_{X_3}^{-1}, \quad J_{14} = A_1(B_{14}^*)^{-1} \\ &J_{21} = A_{24}B_{14}^{-1}, \quad J_{22} = A_2L_{X_2}^{-1}, \quad J_{31} = A_{34}B_{14}^{-1}, \quad J_{33} = A_3L_{X_3}^{-1} \\ &J_{41} = A_4B_{14}^{-1}, \quad J_{56} = A_5(B_{56}^*)^{-1}, \quad J_{65} = A_6B_{56}^{-1}. \end{aligned}$$

Due to $J^* + J = 0$, *we also have* $L_{X_j}A_j + A_j^*L_{X_j} = 0$, $j = 2, 3$, *and*

$$\begin{aligned} &B_{14}^*A_{14} + A_{14}^*B_{14} = 0, \quad L_{X_2}A_{24} + A_{12}^*B_{14} = 0, \quad L_{X_3}A_{34} + A_{13}^*B_{14} = 0 \\ &B_{14}A_4 + A_1^*B_{14} = 0, \qquad B_{56}A_6 + A_5^*B_{56} = 0. \end{aligned}$$

REMARK 2.5. *From the corollary, we have the following observations.*

(i) A_4 *and* $-A_1^*$ *are similar through* B_{14} *and thus have the same spectrum, contained in* $i\mathbf{R}$ *and symmetric about the real axis. This in turn implies that* $\sigma(A_1) = \sigma(A_4)$.

(ii) A_{24} *and* A_{34} *can be determined by other blocks*

$$A_{24} = -L_{X_2}^{-1}A_{12}^*B_{14}, \quad A_{34} = -L_{X_3}^{-1}A_{13}^*B_{14}.$$

Consequently,

$$J_{21} = -L_{X_2}^{-1}A_{12}^*, \quad J_{31} = -L_{X_3}^{-1}A_{13}^*.$$

The proof of Theorem 2.1 is given in Chapter 6, largely based on the Pontryagin invariant subspace theorem 5.1. Theorem 2.1 decomposes the closed operator JL into an upper triangular block form, all of which are bounded except for one block anti-self-adjoint with respective to an equivalent norm. This decomposition plays a fundamental role in proving the linear evolution estimates, the index theorem, the spectral analysis, and the perturbation analysis.

2.3. Exponential Trichotomy

One of our main results is the exponential trichotomy of the semigroup e^{tJL} on X and more regular spaces, to be proved in Chapter 7. Such linear estimates are important for studying nonlinear dynamics, particularly, the construction of invariant manifolds for nonlinear Hamiltonian PDEs.

THEOREM 2.2. *Under assumptions **(H1)**-**(H3)**, JL generates a C^0 group e^{tJL} of bounded linear operators on X and there exists a decomposition*

$$X = E^u \oplus E^c \oplus E^s, \quad \dim E^u = \dim E^s \le n^-(L)$$

satisfying:

i) E^c *and* $E^u, E^s \subset D(JL)$ *are invariant under* e^{tJL}; *Here,* $E^u = X_5$, $E^s = X_6$

are the unstable and stable spaces defined in Theorem 2.1, and the center space E^c *is defined by*

$$E^c = \{u \in X \mid \langle Lu, v\rangle = 0,\ \forall v \in E^s \oplus E^u\} = \oplus_{j=0}^4 X_j;$$

ii) $\langle L\cdot, \cdot\rangle$ *completely vanishes on* $E^{u,s}$*, but is non-degenerate on* $E^u \oplus E^s$*, and*

$$E^c = \{u \in X \mid \langle Lu, v\rangle = 0,\ \forall v \in E^s \oplus E^u\};$$

iii) let $\lambda_u = \min\{Re\lambda \mid \lambda \in \sigma(JL),\ Re\lambda > 0\}$*, there exist* $M > 0$ *and an integer* $k_0 \geq 0$*, such that*

$$\begin{aligned} &\left|e^{tJL}|_{E^s}\right| \leq M(1 + t^{\dim E^s - 1})e^{-\lambda_u t}, \quad \forall\, t \geq 0; \\ &|e^{tJL}|_{E^u}| \leq M(1 + |t|^{\dim E^u - 1})e^{\lambda_u t}, \quad \forall\, t \leq 0, \end{aligned} \tag{2.7}$$

$$|e^{tJL}|_{E^c}| \leq M(1 + |t|^{k_0}), \quad \forall\, t \in \mathbf{R}, \tag{2.8}$$

and

$$k_0 \leq 1 + 2\big(n^-(L) - \dim E^u\big);$$

Moreover, for $k \geq 1$*, define the space* $X^k \subset X$ *to be*

$$X^k = D\big((JL)^k\big) = \{u \in X \mid\ (JL)^n\, u \in X,\ n = 1, \cdots, k.\}$$

and

$$\|u\|_{X^k} = \|u\| + \|JLu\| + \cdots + \|(JL)^k u\|. \tag{2.9}$$

Assume $E^{u,s} \subset X^k$*, then the exponential trichotomy for* X^k *holds true:* X^k *is decomposed as a direct sum*

$$X^k = E^u \oplus E_k^c \oplus E^s,\ E_k^c = E^c \cap X^k$$

and the estimates (2.7) and (2.8) still hold in the norm X^k*.*

An immediate corollary of the theorem is that there are only finitely many eigenvalues of JL outside the imaginary axis in the complex plane.

REMARK 2.6. *The above growth estimates is optimal as one may easily construct finite dimensional examples which achieve upper bounds in the estimates.*

REMARK 2.7. *Naturally, the above invariant decomposition and exponential trichotomy are based on the spectral decomposition of* JL*. The unstable/stable subspaces* $E^{u,s}$ *are the eigenspaces of the stable/unstable spectrum, which have finite total dimensions. Therefore it is easy to obtain the exponential decay estimates of* $e^{tJL}|_{E^{u,s}}$*. While* E^c *is the eigenspace of the spectrum residing on the imaginary axis, the growth estimate of* $e^{tJL}|_{E^c}$ *is far from obvious as the spectral mapping is often a complicated issue especially when continuous spectra is involved. Normally some sub-exponential growth estimates, like in the form of*

$$\forall \epsilon > 0,\ \exists\, C > 0 \Longrightarrow |\ e^{tJL}|_{E^c}\ | \leq Ce^{\epsilon|t|},\ \forall t \in \mathbf{R},$$

are already sufficient for some nonlinear local analysis. Our above polynomial growth estimate on $e^{tJL}|_{E^c}$ *with uniform bound on the degree of the polynomial based on* $\dim X_-$ *is a much stronger statement.*

REMARK 2.8. *Often the invariant subspaces $E^{u,s,c}$ are defined via spectral decompositions where the L-orthogonality between $E^s \oplus E^u$ and E^c is not immediately clear. In fact, this is a special case of more general L-orthogonality property. See Lemma* 6.2 *and Corollary* 6.2.

2.4. Index Theorems and spectral properties

Roughly our next main result is on the relationship between the number of negative directions of L (the Morse index) and the dimensions of various eigenspaces of JL, which may have some implications on $\dim E^{u,s}$ and thus the stability/instability of the group e^{tJL}.

We first introduce some notations. Given any subspace $S \subset X$, denote $n^-(L|_S)$ and $n^{\leq 0}(L|_S)$ as the maximal negative and non-positive dimensions of $\langle Lu, u\rangle$ restricted to S, respectively. Clearly $n^-(L|_s) \leq n^-(L) < \infty$.

In order to state and prove our results on the index theorems, we will work with the standard *complexified* spaces, operators, and quadratic forms, see Appendix for details.

For any eigenvalue λ of JL let E_λ be the generalized eigenspace, that is,

$$E_\lambda = \{u \in X \mid (JL - \lambda I)^k u = 0, \text{ for some integer } k \geq 1\}.$$

REMARK 2.9. *As JL generates a C^0 semigroup (Proposition* A.1*), $(JL-\lambda)^k$ is a densely defined closed operator (see* [**33**]*) and thus E_λ is indeed a closed subspace. It will turn out that $E_\lambda = \ker(JL-\lambda I)^{2n^-(L)+1}$ for any eigenvalue λ. See Theorem* 2.3 *for $\lambda \notin i\mathbf{R}$ and Proposition* 2.1 *for more details.*

Let k_r be the sum of algebraic multiplicities of positive eigenvalues of JL and k_c be the sum of algebraic multiplicities of eigenvalues of JL in the first quadrant (i.e. both real and imaginary parts are positive). Namely,

$$k_r = \sum_{\lambda>0} \dim E_\lambda, \quad k_c = \sum_{Re\lambda,\ Im\lambda>0} \dim E_\lambda. \tag{2.10}$$

For any purely imaginary eigenvalue $i\mu$ ($0 \neq \mu \in \mathbf{R}^+$) of JL, let

$$k^{\leq 0}(i\mu) = n^{\leq 0}\left(L|_{E_{i\mu}}\right), \quad k_i^{\leq 0} = \sum_{0\neq\mu\in\mathbf{R}^+} k^{\leq 0}(i\mu). \tag{2.11}$$

The index counting on E_0 is slightly more subtle due to the possible presence of nontrivial $\ker L \subset E_0$. Observe that, for any subspace $S \subset X$, L induces a quadratic form $\langle L\cdot,\cdot\rangle$ on the quotient space $S/(\ker L \cap S)$. As $\ker L \subset E_0$, define

$$k_0^{\leq 0} = n^{\leq 0}\left(\langle L\cdot,\cdot\rangle|_{E_0/\ker L}\right). \tag{2.12}$$

Equivalently, let $\tilde{E}_0 \subset E_0$ be any subspace satisfying $E_0 = \ker L \oplus \tilde{E}_0$. Define

$$k_0^{\leq 0} = n^{\leq 0}\left(L|_{\tilde{E}_0}\right).$$

It is easy to see that $k_0^{\leq 0}$ is independent of the choice of $\tilde{E}_0$. We have the following index formula which is proved in Section 8.1.

THEOREM 2.3. *Assume* ***(H1)***-***(H3)****, we have*

(i) If $\lambda \in \sigma(JL)$, then $\pm\lambda, \pm\bar{\lambda} \in \sigma(JL)$.

(ii) If λ is an eigenvalue of JL, then $\pm\lambda, \pm\bar{\lambda}$ are all eigenvalues of JL. Moreover, for any integer $k > 0$,

$$\dim \ker(JL \pm \lambda)^k = \dim \ker(JL \pm \bar{\lambda})^k.$$

(iii) The indices satisfy

$$k_r + 2k_c + 2k_i^{\leq 0} + k_0^{\leq 0} = n^-(L)\,. \tag{2.13}$$

Combining Theorem 2.2 and 2.3, we have the following corollary.

COROLLARY 2.2. *(i) If $k_0^{\leq 0} = n^-(L)$, then (2.1) is spectrally stable. That is, there exists no exponentially unstable solution of (2.1).*

(ii) If $n^-(L) - k_0^{\leq 0}$ is odd, then there exists a positive eigenvalue of (2.1), that is, $k_r > 0$. In particular, if $n^-(L) - k_0^{\leq 0} = 1$, then $k_r = 1$ and $k_c = k_i^{\leq 0} = 0$, that is, (2.1) has exactly one pair of stable and unstable simple eigenvalues.

REMARK 2.10. *The formula (2.13) might seem more intuitive if those above $k^{\leq 0}$ had been replaced by k^-. In fact such an index formula with k^- instead of $k^{\leq 0}$ is true only if the quadratic form $\langle Lu, v\rangle$ is non-degenerate on all $E_{i\mu}$, $\mu \in \mathbf{R}^+$ and $\tilde{E}_0$, which would imply $n^-(L|_{E_{i\mu}}) = n^{\leq 0}(L|_{E_{i\mu}})$. However, the degeneracy is indeed possible and the correct choice has to be $k^{\leq 0}$. Such an example is given in Section 8.4.*

Even though we can not claim $\dim E_{i\mu} < \infty$ for an eigenvalue $i\mu \in i\mathbf{R}$ which might be embedded in the continuous spectrum, in fact $E_{i\mu}$ is spanned by eigenvectors along with finitely many generalized eigenvectors, except for $\mu = 0$. More precisely, we prove the following two propositions in Lemma 3.5 and Section 8.2.

PROPOSITION 2.1. *Assume **(H1)**-**(H3)**. For any $\mu \in \sigma(JL) \cap i\mathbf{R}\backslash\{0\}$, it holds*

$$E_{i\mu} = \ker(JL - i\mu)^{2k^{\leq 0}(i\mu)+1}, \quad \dim\left((JL - i\mu)E_{i\mu}\right) \leq 2k^{\leq 0}(i\mu).$$

Moreover,

$$E_0 = \ker(JL)^{2k_0^{\leq 0}+2}, \quad \dim\left((JL)^2 E_0\right) \leq 2k_0^{\leq 0}.$$

The above proposition does not hold if $(JL)^2 E_0$ is replaced by JLE_0 as in the case of $\mu \neq 0$. See an example in Remark 8.2 in Section 8.2.

For $\mu \in \mathbf{R}$, Theorem 2.3 and Proposition 2.1 mean that, in addition to eigenvectors, $JL|_{E_{i\mu}}$ has only finitely many nontrivial Jordan blocks with the total dimensions bounded in term of $n^-(L)$. The number and the lengths of nontrivial Jordan chains of $JL|_{E_{i\mu}}$ are independent of the choice of the basis realizing the Jordan canonical form. Intuitively if a basis consisting of generalized eigenvectors simultaneously diagonalizes the quadratic form $\langle Lu, u\rangle$ and realizes the Jordan canonical form of JL, it would greatly help us to understand the structure of (2.1). However, usually this is not possible. Instead, we find a 'good' basis for the Jordan canonical form of JL which also 'almost' diagonalizes the quadratic form L. To our best knowledge, we are not aware of such a result even in finite dimensions.

PROPOSITION 2.2. *Assume **(H1)**-**(H3)**. For $i\mu \in \sigma(JL) \cap i\mathbf{R}\backslash\{0\}$, there exists a decomposition of $E_{i\mu}$ into closed subspaces $E_{i\mu} = E^D \oplus E^1 \oplus E^G$ such that L and JL take the block forms*

$$\langle L\cdot, \cdot\rangle \longleftrightarrow \begin{pmatrix} 0 & 0 & 0 \\ 0 & L_1 & 0 \\ 0 & 0 & L_G \end{pmatrix}, \quad JL \longleftrightarrow \begin{pmatrix} A_D & A_{D1} & A_{DG} \\ 0 & i\mu & 0 \\ 0 & 0 & A_G \end{pmatrix}.$$

For $\mu = 0$, there exists a decomposition $E_0 = \ker L \oplus E^D \oplus E^1 \oplus E^G$ such that L and JL take the block form

$$\langle L\cdot,\cdot\rangle \longleftrightarrow \begin{pmatrix} 0 & 0 & 0 & 0 \\ 0 & 0 & 0 & 0 \\ 0 & 0 & L_1 & 0 \\ 0 & 0 & 0 & L_G \end{pmatrix}, \quad JL \longleftrightarrow \begin{pmatrix} 0 & A_{0D} & A_{01} & A_{0G} \\ 0 & A_D & A_{D1} & A_{DG} \\ 0 & 0 & 0 & 0 \\ 0 & 0 & 0 & A_G \end{pmatrix}.$$

In both cases, all blocks are bounded operators, L_1 and L_G are non-degenerate, $\sigma(A_G) = \sigma(A_D) = \{i\mu\}$, and

$$\dim E^G \leq 3\big(k^{\leq 0}(i\mu) - \dim E^D - n^-(L|_{E^1})\big), \quad \dim E^1 \leq \infty.$$

Moreover, $\ker(A_G - i\mu) \subset (A_G - i\mu)E^G$, namely, the Jordan canonical form of JL on E^G has non-trivial blocks only. Let $1 < k_1 < \cdots < k_{j_0}$ be the dimensions of Jordan blocks of A_G in E^G. Suppose there are l_j Jordan blocks of size $k_j \times k_j$. For each $j = 1, \ldots, j_0$, there exist linearly independent vectors

$$\{u_{p,q}^{(j)} \mid p = 1, \ldots, l_j,\ q = 1, \ldots, k_j\} \subset E^G \tag{2.14}$$

such that

(1) $\forall\ 1 \leq p \leq l_j$,

$$\left\{u_{p,q}^{(j)} = (JL - i\mu)^{q-1} u_{p,1}^{(j)},\ q = 1, \ldots, k_j\right\}$$

form a Jordan chain of length k_j. More explicitly ,

on span $\{u_{1,1}^{(j)}, \ldots, u_{1,k_j}^{(j)}, \ldots, u_{l_j,1}^{(j)}, \ldots, u_{l_j,k_j}^{(j)}\}$:

$$A_G \longleftrightarrow \begin{pmatrix} i\mu & 0 & \cdots & 0 & 0 & \cdots & 0 & 0 & \cdots & 0 & 0 \\ 1 & i\mu & \cdots & 0 & 0 & \cdots & 0 & 0 & \cdots & 0 & 0 \\ & & \cdots & & & & & & & & \\ 0 & 0 & \cdots & 1 & i\mu & \cdots & 0 & 0 & \cdots & 0 & 0 \\ & & & & \cdots & & & & & & \\ 0 & 0 & \cdots & 0 & 0 & \cdots & i\mu & 0 & \cdots & 0 & 0 \\ 0 & 0 & \cdots & 0 & 0 & \cdots & 1 & i\mu & \cdots & 0 & 0 \\ & & & & & & & \cdots & & & \\ 0 & 0 & \cdots & 0 & 0 & \cdots & 0 & 0 & \cdots & 1 & i\mu \end{pmatrix}$$

The above count for all Jordan blocks of A_G of size k_j.

(2) $\langle Lu_{p,q}^{(j)}, u_{p',q'}^{(j')}\rangle = 0$ *if $p \neq p'$ or $j \neq j'$.*

(3) $\forall\ 1 \leq p \leq l_j$, *the $k_j \times k_j$ representation matrix of L on a chain* (2.14) *is*

$$\big(\langle Lu_{p,q}^{(j)}, u_{p,r}^{(j)}\rangle\big)_{q,r} = \begin{pmatrix} 0 & 0 & \cdots & 0 & a_{p,1}^{(j)} \\ 0 & 0 & \cdots & a_{p,2}^{(j)} & 0 \\ \cdots & & & & \\ a_{p,k_j}^{(j)} & 0 & \cdots & 0 & 0 \end{pmatrix}, \tag{2.15}$$

where the entries satisfy

$$a_{p,q'}^{(j)} = (-1)^{q'-q} a_{p,q}^{(j)} \neq 0, \qquad a_{p,k_j+1-q}^{(j)} = \overline{a_{p,q}^{(j)}}$$

and thus the above matrix is non-degenerate.

(4) *If k_j is odd, then* $a^{(j)}_{p,\frac{1}{2}(k_j+1)} = \pm 1$ *and the k_j-th Krein signature of $i\mu$ defined by*

$$n^-_{k_j}(i\mu) = \sum_{p=1}^{l_j} \min\{0,\ a^{(j)}_{p,\frac{1}{2}(k_j+1)}\}$$

is independent of the choices of such bases $\{u^{(j)}_{p,q}\}$.

REMARK 2.11. *Since $\langle Lu, u\rangle$ is symmetric (Hermitian after the complexification), we can normalize the above $a^{(j)}_{p,q}$ such that $a^{(j)}_{p,q} = \pm 1$ if k_j is odd and $a^{(j)}_{p,q} = \pm i$ if k_j is even. In particular, when $\mu = 0$, since the generalized eigenspace is spanned by real functions in X, it follows that the Jordan chains in $E^G \subset E_0$ are all of odd length.*

In the splitting of $E_{i\mu}$, we note that only E^1 may be infinite dimensional, where L is positive except in finitely many directions. If $\langle L\cdot, \cdot\rangle$ is non-degenerate on $E_{i\mu}$, the subspace E^D may be eliminated and many of our results can be improved. However, this degeneracy indeed is possible. See such an example in Section 8.4. On the positive side, in that section, we also prove the following proposition on the non-degeneracy of $L|_{E_{i\mu}}$ for isolated eigenvalues $i\mu$. In particular, the isolation assumption for $i\mu \in \sigma(JL) \cap i\mathbf{R}$ usually holds if the problem comes from PDEs defined on bounded or periodic domains.

PROPOSITION 2.3. *If $i\mu \in \sigma(JL) \cap i\mathbf{R}$ is isolated in $\sigma(JL)$, then*

(i) $i\mu$ is an eigenvalue, i.e. $E_{i\mu} \neq \{0\}$, and $\langle L\cdot, \cdot\rangle$ is non-degenerate on $E_{i\mu}/(\ker L \cap E_{i\mu})$.

(ii) there exists a closed subspace $E_\# \subset X$ invariant under JL such that $X = E_{i\mu} \oplus E_\#$ and $\langle Lu, v\rangle = 0$ for all $u \in E_{i\mu}$ and $v \in E_\#$.

(iii) $\sigma\big((JL)|_{E_\#}\big) = \sigma(JL)\backslash\{i\mu\}$.

In the case of an isolated spectral point $i\mu$, one may define the invariant eigenspaces and its complement eigenspace via contour integral in operator calculus. Usually it is not guaranteed that such $i\mu$ is an eigenvalue and its eigenspace coincides with $E_{i\mu}$. This proposition implies that, under assumptions (**H1-3**), this is exactly the case and $\langle L\cdot, \cdot\rangle$ is non-degenerate on $E_{i\mu}$. As a corollary, we prove

PROPOSITION 2.4. *In addition to (**H1-3**), we assume*

(**H4**) *$\langle L\cdot, \cdot\rangle$ is non-degenerate on E_λ for any non-isolated $\lambda \in \sigma(JL) \cap i\mathbf{R}\backslash\{0\}$ and also on $E_0/\ker L$ if $0 \in \sigma(JL)$ is not isolated,*

then there exist closed subspaces N and M, which are L-orthogonal, such that $N \oplus \ker L$ and $M \oplus \ker L$ are invariant under JL, $X = N \oplus M \oplus \ker L$, $\dim N < \infty$, and $L \geq \delta$ on M for some $\delta > 0$.

In particular, if eigenvalues of JL are isolated, then by Proposition 2.3, (**H4**) is automatically satisfied and Proposition 2.4 holds. If we further assume $\ker L = \{0\}$, then $X = N \oplus M$ and both N and M are invariant under JL. Proposition 2.4 can be used to construct invariant decompositions for $L-$self-adjoint operators. The next proposition gives a generalization of Theorem A.1 in [**23**], which was proved for a compact L-self-adjoint operator A with $\ker A = \{0\}$. Such decomposition was used to study the damping of internal waves in a stably stratified fluid ([**23**]).

PROPOSITION 2.5. *Let X be a complex Hilbert space along with a Hermitian symmetric quadratic form $B(u,v) = \langle Lv, u\rangle$ defined by an (anti-linear) operator $L : X \to X^*$ satisfying **(H2)** with* $\ker L = \{0\}$. *Let $A : X \to X$ be a $L-$self-adjoint complex linear operator (i.e. $\langle LAu, v\rangle = \langle Lu, Av\rangle$) such that nonzero eigenvalues of A are isolated. If $L|_{\ker A}$ is non-degenerate, then there exists a decomposition $X = N \oplus M$ such that N and M are L-orthogonal and invariant under A,* $\dim N < \infty$ *and $L|_M$ is uniformly positive.*

We will extend the notion of the *Krein signature* to eigenvalues $i\mu$ for which $\langle L\cdot,\cdot\rangle$ on $E_{i\mu}$ is non-degenerate, and give more detailed descriptions of k_i^- and k_0^-. As commented above, the non-degeneracy assumption means E^D is eliminated in $E_{i\mu}$. For such μ, define

$$E_{i\mu,0} = \{v \in \ker(JL - i\mu) \mid \langle Lv, u_{p,q}^{(j)}\rangle = 0,\ \forall 1 \le j \le j_0,\ 1 \le p \le l_j,\ 1 \le q \le k_j\}$$

which is the complementary subspace of $R(JL - i\mu) \cap \ker(JL - i\mu)$ inside $\ker(JL - i\mu)$. It corresponds to the diagonalized part of $JL|_{E_{i\mu}}$.

DEFINITION 2.1. *For $\mu \ge 0$ such that $\langle L\cdot,\cdot\rangle$ is non-degenerate on $E_{i\mu}$, define the first Krein signature*

$$n_1^-(i\mu) = n^-(L|_{E_{i\mu,0}})$$

and k_j-th Krein signatures as $n_{k_j}^-(i\mu)$ given in Proposition 2.2, for odd $k_j = 2m - 1 \ge 1$.

REMARK 2.12. *The Krein signature $n_{k_j}^-(i\mu)$, for odd $k_j = 2m-1 \ge 1$, does not have to be defined as in Proposition 2.2 using the above special bases. In fact, for any j, let $\left\{v_{p,q}^{(j)}\right\}$ be an arbitrary complete set of Jordan chains of length k_j. Define the $l_j \times l_j$ matrix $\tilde{M}_j = \left(\left\langle Lv_{p_1,m}^{(j)}, v_{p_2,m}^{(j)}\right\rangle\right)$, $1 \le p_1, p_2 \le l_j$. Then $n_{k_j}^-(i\mu) = n^-\left(\tilde{M}_j\right)$, the negative index (Morse index) of $\tilde{M}_j$.*

REMARK 2.13. *The signatures $n_{k_j}^-(\mu)$ may also be defined in an intrinsic way independent of bases. See Definition 4.1 and equation (4.2).*

According to Proposition 2.2, the 2-dim subspace $\mathrm{span}\{u_{p,q}^{(j)}, u_{p,k_j+1-q}^{(j)}\}$ and 1-dim subspace $\mathrm{span}\{u_{p,\frac{1}{2}(k_j+1)}^{(j)}\}$ for odd k_j are L-orthogonal to each other. With respect to the basis $\{u_{p,q}^{(j)}, u_{p,k_j+1-p}^{(j)}\}$ there, L takes the form of the Hermitian symmetric matrix $\begin{pmatrix} 0 & a \\ \bar{a} & 0 \end{pmatrix}$ with $a \neq 0$, whose Morse index is clearly 1. Therefore we obtain the following formula for k_i^-.

PROPOSITION 2.6. *In addition to **(H1)**-**(H3)**, assume $i\mu \in \sigma(JL) \cap i\mathbf{R}$ satisfies that $\langle L\cdot,\cdot\rangle$ is non-degenerate on $E_{i\mu}$. Then we have*

$$k^{\le 0}(i\mu) = k^-(i\mu) = \sum_{k_j\ even} \frac{l_j k_j}{2} + \sum_{k_j\ odd} \left[\frac{l_j(k_j - 1)}{2} + n_{k_j}^-(i\mu)\right]. \tag{2.16}$$

As Hamiltonian systems often possess additional symmetries which generate nontrivial $\ker L$, $k_0^{\le 0}$ deserves some more discussion if $\ker L \neq \{0\}$. The following propositions are proved in Section 8.3, based on a decomposition of the subspace E_0. Recall that for any subspace $S \subset X$, L also induces a quadratic form $\langle L\cdot,\cdot\rangle$ on the quotient space $S/(S \cap \ker L)$.

PROPOSITION 2.7. *Assume* ***(H1)-(H3)****, then* $(JL)^{-1}(\ker L)$ *is a closed subspace. Furthermore, let*

$$n_0 = n^{\leq 0}(\langle L\cdot,\cdot\rangle|_{(JL)^{-1}(\ker L)/\ker L}).$$

Then

(i) $k_0^{\leq 0} \geq n_0$.

(ii) If $\langle L\cdot,\cdot\rangle$ *is non-degenerate on* $(JL)^{-1}(\ker L)/\ker L$*, then*

$$k_0^{\leq 0} = n_0 = n^{-}(\langle L\cdot,\cdot\rangle|_{(JL)^{-1}(\ker L)/\ker L}).$$

REMARK 2.14. *Practically, in order to compute* n_0 *in the above proposition, let* $S \subset (JL)^{-1}(\ker L)$ *be a closed subspace such that*

$$(JL)^{-1}(\ker L) = \ker L \oplus S, \tag{2.17}$$

then $n_0 = n^{\leq 0}(L|_S)$*. Often* S *can be taken as* $(\ker L)^{\perp} \cap (JL)^{-1}(\ker L)$.

It is worth comparing the above results with some classical results (e.g. [**29**, **30**]). Consider a nonlinear Hamiltonian equation

$$\partial_t u = JDH(u) \tag{2.18}$$

which has an additional conserved quantity $P(u)$ (often the momentum, mass *etc.*) due to some symmetry. Assume that for c in a neighborhood of c_0, there exists u_c such that $DH(u_c) - cDP(u_c) = 0$, which gives a relative equilibrium of (2.18) such as traveling waves, standing waves, etc. The linearized equation of (2.18) in some reference frame at u_{c_0} takes the form of (2.1) with $L = D^2H(u_{c_0}) - c_0D^2P(u_{c_0})$. It can be verified that $JDP(u_{c_0}) \in \ker L$ and $L\partial_c u_c|_{c=c_0} = DP(u_{c_0})$. In the case where $\ker L = span\{JDP(u_{c_0})\}$ and J is one to one (not necessarily with bounded J^{-1} as assumed in [**29**, **30**]), we have

$$(JL)^{-1}(\ker L) = span\{JDP(u_{c_0}), \partial_c u_c|_{c=c_0}\}$$

when $\frac{d}{dc}P(u_c)|_{c=c_0} \neq 0$ and

$$n_0 = \begin{cases} 0 & \text{if } \frac{d}{dc}P(u_c)|_{c=c_0} < 0 \\ 1 & \text{if } \frac{d}{dc}P(u_c)|_{c=c_0} > 0 \end{cases}.$$

If we further assume $n^{-}(L) = 1$, then the combination of Proposition 2.7 and Theorem 2.3 implies the result in [**29**] that equation (2.1) is stable if $\frac{d}{dc}P(u_c)|_{c=c_0} \leq 0$ and unstable if $\frac{d}{dc}P(u_c)|_{c=c_0} > 0$.

In the following special cases, $k_0^{\leq 0}$ as well as n_0 can be better estimated, which is often useful in applications.

LEMMA 2.1. *Assume* ***(H1)-(H3)****. we have*

(i) $\langle Lu, v\rangle = 0, \quad \forall u \in \ker(JL),\ v \in \overline{R(J)}$.

(ii) $\langle Lu, u\rangle$ *is non-degenerate on* $\ker(JL)/\ker L$ *if and only if it is non-degenerate on* $\overline{R(J)}/\big(\ker L \cap \overline{R(J)}\big)$.

While the statement of the lemma and the following proposition in the language of quotient spaces make them independent of choices of subspaces transversal to $\ker L$, practically it might be easier to work with subspaces. The following is an equivalent restatement of Lemma 2.1 using subspaces. Actually the proof in Section 8.3 will be carried out by using subspaces.

COROLLARY 2.3. *Let $S_1, S^{\#} \subset X$ be closed subspaces such that*

$$\ker(JL) = \ker L \oplus S_1, \qquad \overline{R(J)} = \left(\overline{R(J)} \cap \ker L\right) \oplus S^{\#}. \tag{2.19}$$

We have that $\langle L\cdot, \cdot\rangle$ is non-degenerate on S_1 if and only if it is non-degenerate on $S^{\#}$.

Under this non-degeneracy, we have

PROPOSITION 2.8. *Assume **(H1)**-**(H3)**, and that $\langle Lu, u\rangle$ is non-degenerate on $\ker(JL)/\ker L$ which is equivalent to $\ker(JL) \cap \overline{R(J)} \subset \ker L$, then*

(i) $X = \ker(JL) + \overline{R(J)}$ and

$$n^-(L) = n^-\left(L|_{\ker(JL)/\ker L}\right) + n^-\left(L|_{\overline{R(J)}/\left(\ker L \cap \overline{R(J)}\right)}\right).$$

(ii) Let

$$\tilde{S} = \overline{R(J)} \cap (JL)^{-1}(\ker L).$$

Then

$$k_0^{\leq 0} \geq n^-(L|_{\ker(JL)/\ker L}) + n^{\leq 0}(L|_{\tilde{S}/(\ker L \cap \tilde{S})}).$$

(iii) If, in addition, $\langle Lu, u\rangle$ is non-degenerate on $\tilde{S}/(\ker L \cap \tilde{S})$, then

$$\begin{aligned} k_0^{\leq 0} &= n^-(L|_{\ker(JL)/\ker L}) + n^-(L|_{\tilde{S}/(\ker L \cap \tilde{S})}) \\ k_r + 2k_c + 2k_i^{\leq 0} &= n^-\left(L|_{\overline{R(J)}/\left(\ker L \cap \overline{R(J)}\right)}\right) - n^-\left(L|_{\tilde{S}/(\ker L \cap \tilde{S})}\right). \end{aligned} \tag{2.20}$$

We notice that the last equality is only a consequence of the previous two equalities on n^- and $k_0^{\leq 0}$ and the index Theorem 2.3.

In terms of subspaces, equivalently we have

COROLLARY 2.4. *Let $S_1, S^{\#} \subset X$ be closed subspaces assumed in Corollary 2.3 and $S_2 \in X$ be a closed subspace such that*

$$\overline{R(J)} \cap (JL)^{-1}(\ker L) = S_2 \oplus \left(\overline{R(J)} \cap \ker L\right). \tag{2.21}$$

Assume the non-degeneracy of $\langle Lu, u\rangle$ on S_1. Under this condition, we have

$$X = \ker L \oplus S_1 \oplus S^{\#},$$

and this decomposition is orthogonal with respect to the quadratic form $\langle L\cdot, \cdot\rangle$. Moreover, we have

$$n^-(L) = n^-(L_{S_1}) + n^-(L|_{S^{\#}}) \ \text{ and } \ k_0^{\leq 0} \geq n^-(L|_{S_1}) + n^{\leq 0}(L|_{S_2}).$$

The additional non-degeneracy assumption of $\langle Lu, u\rangle$ on $\tilde{S}/(\ker L \cap \tilde{S})$ is equivalent to its non-degeneracy on S_2 and it implies

$$\begin{aligned} k_0^- &= n^-(L|_{S_1}) + n^-(L|_{S_2}) \\ k_r + 2k_c + 2k_i^- &= n^-\left(L|_{S^{\#}}\right) - n^-(L|_{S_2}). \end{aligned}$$

Very often subspaces $S_1, S^{\#}, S_2$ can be taken as various intersections with $(\ker L)^{\perp}$.

2.5. Structural stability/instability

Our next main result is on the spectral properties of the Hamiltonian operator JL under small bounded perturbations. Consider the perturbed linear Hamiltonian system

$$u_t = J_\# L_\# u, \qquad J_\# = J + J_1, \quad L_\# = L + L_1, \quad u \in X. \tag{2.22}$$

We assume the perturbations satisfy

- (**A1**) J and L satisfies (**H1-2**) and the perturbations $J_1 : X^* \to X$ and $L_1 : X \to X^*$ are bounded operators with $J_1^* = -J_1$ and $L_1^* = L_1$.
- (**A2**) $\dim \ker L < \infty$;
- (**A3**) $D(JL) \subset D(JL_1)$.

We note that (**A2**) implies (**H3**) for JL by Remark 2.3. From the Closed Graph Theorem, JL_1 is a bounded operator on the Hilbert space $D(JL)$ equipped with the graph norm

$$||u||_G^2 \triangleq ||u||^2 + ||JLu||^2, \ u \in D(JL); \ |JL_1|_G \triangleq \sup_{||u||_G=1} ||JL_1 u||. \tag{2.23}$$

We first point out that assumptions (**A1-3**) imply (**H1-3**) for $J_\# L_\#$ when the perturbations are sufficiently small as assumed in Theorem 2.4 below. See Lemma 9.1. As indicated in assumption (**A1**) we consider bounded perturbations to both the symplectic structure J and the energy quadratic form L, while the Hamiltonian structure is preserved. Assumption (**A2**) ensures $n^-(L_\#) < \infty$ so that the perturbed problem is still in our framework. Assumption (**A3**) is a regularity assumption which implies that $J_\# L_\#$ is not more unbounded compared to JL. Therefore the resolvent $(\lambda - J_\# L_\#)^{-1}$ is a small perturbation of $(\lambda - JL)^{-1}$ as proved in Lemma 9.2.

Let $E^{u,s,c}$ be the unstable/stable/center subspaces of JL, as well as the constants $\lambda_u > 0$, as given in Theorem 2.2. The next theorem and the following proposition will be proved in Section 9.1.

THEOREM 2.4. *Assume (A1-3). There exist $C, \epsilon_0 > 0$ depending only on J and L such that, if*

$$|J_1| + |L_1| + |JL_1|_G \le \epsilon \le \epsilon_0, \tag{2.24}$$

then

(a) *There exist bounded operators*

$$S_\#^u : E^u \to E^s \oplus E^c, \quad S_\#^s : E^s \to E^u \oplus E^c, \quad S_\#^c : E^c \to E^s \oplus E^u,$$

such that

$$|S_\#^{u,s,c}| \le C\epsilon, \quad e^{tJ_\# L_\#} E_\#^{u,s,c} = E_\#^{u,s,c}, \quad \textit{where } E_\#^{u,s,c} = \mathrm{graph}(S_\#^{u,s,c}),$$

for all $t \in \mathbf{R}$. Moreover,

$$\begin{aligned} \left| e^{tJ_\# L_\#}|_{E_\#^s} \right| &\le C(1 + t^{\dim E^s - 1}) e^{-(\lambda_u - C\epsilon)t}, \quad \forall\, t \ge 0; \\ |e^{tJ_\# L_\#}|_{E_\#^u}| &\le C(1 + |t|^{\dim E^u - 1}) e^{(\lambda_u - C\epsilon)t}, \quad \forall\, t \le 0, \end{aligned} \tag{2.25}$$

$$|e^{tJ_\# L_\#}|_{E_\#^c}| \le C\epsilon^{\frac{1}{2(1+n^-(L) - \dim E^u)} - 1} e^{C\epsilon^{\frac{1}{2(1+n^-(L) - \dim E^u)}} |t|}, \quad \forall\, t \in \mathbf{R}. \tag{2.26}$$

(b) $\langle L_\# \cdot, \cdot\rangle$ *vanishes on* $E_\#^{u,s}$*, but is non-degenerate on* $E_\#^s \oplus E_\#^u$*, and*

$$E_\#^c = \{u \mid \langle L_\# u, v\rangle = 0, \ \forall v \in E_\#^u \oplus E_\#^s\}.$$

(c) *If* $\langle L\cdot, \cdot\rangle \geq \delta > 0$ *on* E^c*, then there exists* $C' > 0$ *depending on* δ*,* J*, and* L *such that* $|e^{tJ_\# L_\#}|_{E_\#^c}| \leq C'$ *for any* $t \in \mathbf{R}$*.*

Due to assumption (**A3**), the resolvent $(\lambda - J_\# L_\#)^{-1}$ is only a small perturbation of $(\lambda - JL)^{-1}$ as proved in Lemma 9.2. Therefore the existence of the invariant subspaces $E_\#^{u,s,c}$ as a small perturbation to $E^{u,s,c}$ follows immediately. Statements (b) and (c) basically result from the Hamiltonian structure and the estimates of $e^{tJ_\# L_\#}$ on $E_\#^{u,s}$ are basically due to their finite dimensionality. If $J_\# L_\# - JL$ had been a bounded operator, estimate (2.26) would follow easily from the standard spectral theory as well. However, since $J : X^* \supset D(J) \to X$ is only assumed to satisfy $J^* = -J^*$, the term JL_1 may not be bounded and thus (2.26) does not follow from the standard spectral theory. Our proof heavily relies on the decomposition given by Theorem 2.1. In fact, the usual resolvent estimate often neglects the Hamiltonian structure of the problem which actually plays an essential role here. Otherwise a counterexample without the Hamiltonian structure is $J = J_\# = i$ and $L_\# = \partial_{xx} + \epsilon\partial_x$ with $X = H^1(S^1, \mathbf{C})$, for which the equation $u_t = J_\# L_\# u$ is not even well-posed in X for $\epsilon \neq 0$.

Another consequence of Lemma 9.2 of the resolvent estimate and Lemma 6.2 is the following structural stability type result.

Proposition 2.9. *Suppose closed subsets* $\sigma_{1,2} \subset \sigma(JL)$ *satisfy*

(1) $\sigma(JL) = \sigma_1 \cup \sigma_2$*,* $\sigma_1 \cap \sigma_2 = \emptyset$*, and* σ_2 *is compact.*

(2) *For any* $\lambda \in \sigma_1$ *and* $0 \neq u \in E_\lambda$*, it holds* $\langle Lu, u\rangle > 0$*.*

Then there exist $\alpha, \epsilon_0 > 0$ *depending only on* J *and* L *such that* (2.24) *implies*

$$\{\lambda \in \sigma(J_\# L_\#) \mid d(\lambda, \sigma_2) > \alpha\} \subset i\mathbf{R}.$$

From Proposition 6.2, any $\lambda \in \sigma(JL)\backslash i\mathbf{R}$ is an eigenvalue, i.e. $E_\lambda \neq \{0\}$, and $\langle L\cdot, \cdot\rangle$ vanishes on E_λ. Therefore it must hold that $\sigma_1 \subset i\mathbf{R}$. Even though the second assumption on σ_1 seems weaker than that $\langle L\cdot, \cdot\rangle$ is uniformly positive on its eigenspaces, it along with Theorem 2.3 actually implies the latter. This proposition means that, under small perturbations, unstable eigenvalues can not bifurcate from such σ_1.

In the next we consider the deformation of purely imaginary spectral points of JL under perturbations as they are closely related to generation of linear instability. The next two theorems are proved in Section 9.2. Firstly we prove that if $i\mu \in \sigma(JL)$ and $\langle L\cdot, \cdot\rangle$ has certain definite sign on $E_{i\mu}$, then $\sigma(J_\# L_\#)$ would not have nearby unstable eigenvalues.

Theorem 2.5. *Assume (A1-3),* $i\mu \in \sigma(JL) \cap i\mathbf{R}$*, and either a.) there exists* $\delta > 0$ *such that* $\langle Lu, u\rangle \geq \delta \|u\|^2$ *for all* $u \in E_{i\mu}$ *or b.)* $i\mu$ *is isolated in* $\sigma(JL)$*and* $\langle Lu, u\rangle \leq -\delta\|u\|^2$ *for all* $u \in E_{i\mu}$*, then there exist* $\alpha, \epsilon_0 > 0$ *depending on* J*,* L*,* μ*, and* δ *such that, if* (2.24) *holds, then*

$$\{\lambda \in \sigma(J_\# L_\#) \mid |\lambda - i\mu| \leq \alpha\} \subset i\mathbf{R}.$$

Remark 2.15. *On the one hand, note that in the above theorem, we do not require* $i\mu$ *being an isolated eigenvalue or even an eigenvalue of* JL*. If* $i\mu$ *is not an eigenvalue,* $E_{i\mu} = \{0\}$ *and the sign definiteness assumption is automatically* *satisfied. On the other hand, if* $i\mu$ *is an isolated spectral point, then Proposition* 2.3 *implies that* $E_{i\mu}$ *is nontrivial and is precisely the eigenspace of* $i\mu$*. Moreover, from Lemma* 3.4 *and the sign definiteness of* L *on* $E_{i\mu}$*, we have* $E_{i\mu} = \ker(JL - i\mu)$.

On the one hand, the above theorem indicates that under Hamiltonian perturbations, hyperbolic (i.e. stable and unstable) eigenvalues can not bifurcate from either a.) any $i\mu \in \sigma(JL)$, whether isolated or not, for which $\langle L\cdot, \cdot\rangle$ is positive on $E_{i\mu}$, or b.) any isolated eigenvalue $i\mu$ where $\langle L\cdot, \cdot\rangle$ has a definite sign on $E_{i\mu}$. Theorem 2.5, as well as Theorem 2.4 can be viewed as robustness or structural stability type results.

On the other hand, as given in the next theorem, the structural stability conditions in Theorem 2.5 are also necessary for an eigenvalue $i\mu \neq 0$. As in many applications parameters mostly appear in the energy operator L instead of the symplectic operator J, we will study perturbations only to L for possible bifurcations of unstable eigenvalues near $i\mu$.

THEOREM 2.6. *Assume that* (J, L) *satisfies* ***(H1-3)*** *and* $0 \neq i\mu \in \sigma(JL) \cap i\mathbf{R}$ *satisfies*

(1) $\langle L\cdot, \cdot\rangle$ *is neither positive nor negative definite on* $E_{i\mu}$ *or*
(2) $i\mu$ *is non-isolated in* $\sigma(JL)$ *and there exists* $u \in E_{i\mu}$ *with* $\langle Lu, u\rangle \leq 0$,

then for any $\epsilon > 0$*, there exist a symmetric bounded linear operator* $L_1 : X \to X^*$ *such that:* $|L_1| < \epsilon$ *and there exists* $\lambda \in \sigma\big(J(L + L_1)\big)$ *with* $\operatorname{Re}\lambda > 0$ *and* $|\lambda - i\mu| < C\epsilon$*, for some constant* C *depending only on* μ, J, L.

It is easy to see that conditions in Theorem 2.6 are exactly complementary to those in Theorem 2.5 for $i\mu \neq 0$ and thus they give necessary and sufficient conditions on whether unstable eigenvalues can bifurcate from $0 \neq i\mu \in \sigma(JL) \cap i\mathbf{R}$ under Hamiltonian perturbations.

REMARK 2.16. *In* [**28**]*, Grillakis proved that an embedded purely imaginary eigenvalue with negative energy of the linearized operator at excited states of a semilinear nonlinear Schrödinger equation is 'structurally unstable' under small perturbations and unstable eigenvalues can be generated. The linearized operator is of the form* JL*, where*

$$J = \begin{pmatrix} 0 & 1 \\ -1 & 0 \end{pmatrix}, \ L = \begin{pmatrix} -\Delta + V_1(x) & 0 \\ 0 & -\Delta + V_2(x) \end{pmatrix}.$$

Here, $V_1(x), V_2(x) \to \omega > 0$ *exponentially when* $|x| \to \infty$*. Under some assumptions, Theorem 2.4 in* [**28**] *implies that that if* $i\mu \neq 0$ *is an embedded eigenvalue of* JL *with* $\langle Lu, u\rangle < 0$ *for some eigenfunction* u*, then an unstable eigenvalue may bifurcate from* $i\mu$ *under Hamiltonian perturbations. Similar result was also obtained in* [**21**]*. This is a special case of the above theorem. Actually, we can relax the structural instability condition to be that* $\langle L\cdot, \cdot\rangle$ *is not positive definite on* $E_{i\mu}$*, including cases of degeneracy of* $L|_{E_{i\mu}}$ *or with Jordan chains.*

However, it should be pointed out that it is not clear that the above structural instability may be realized by the linearized equation of the nonlinear Schrödinger equation at a perturbed excited state. It would be interesting to see if one can prove the structural instability in the sense that there is linear instability for nearby excited states.

REMARK 2.17. *The case $i\mu = 0$ is not included in Theorem 2.6 since this may be related to some additional degeneracy of L or J. See for example Cases 3b and 3d in Section 9.2. The analysis of possible bifurcations of unstable eigenvalues from $\mu = 0$ could be carried out in a similar fashion based on the Propositions 2.2, 2.3, Lemma 9.5, etc., but more carefully. We feel that it might be easier to work on this case directly in concrete applications and thus do not include it in the above theorem.*

2.6. A theorem where L does not have a positive lower bound on X_+

Among our global assumptions (**H1-3**), (**H2**) requires that the phase space X is decomposed into the direct sum of three subspaces $X = X_- \oplus \ker L \oplus X_+$, such that the quadratic form $\langle L\cdot, \cdot\rangle$ is uniformly positive/negative on $X_\pm$. This assumption plays a crucial role in the analysis throughout the paper. However, in some Hamiltonian PDEs L, which usually appears as the Hessian of the energy functional at a steady state, may not have a positive lower bound on X_+. One such simple example is $X = H^1(\mathbf{R}^n)$ and $L = -\Delta + a(x)$ where $\lim_{|x|\to\infty} a(x) = 0$. Even if $a > 0$ which implies $L > 0$, but for any $\delta > 0$, there exists $u \in H^1$ such that $\langle Lu, u\rangle < \delta \|u\|^2_{H^1}$. A potential resolution to this issue in this specific example is to take a different phase space such as $\dot{H}^1$ instead of H^1. In Chapter 10, we show that this observation may be applied in a rather general setting. As a non-trivial example of this case, the stability of traveling waves of a nonlinear Schrödinger equation in 2-dim with non-vanishing condition at $|x| = \infty$ is considered in Section 11.6.

In this section, let X be a real Hilbert space with the inner product $(\cdot, \cdot)$ and we assume

(**B1**) $Q_0, Q_1 : X \to X^*$ are bounded positive symmetric linear operators such that

$$\langle (Q_0 + Q_1)u, v\rangle = (u, v), \ Q_{0,1}^* = Q_{0,1}, \ \langle Q_{0,1}u, u\rangle > 0, \ \forall\, 0 \neq u, v \in X.$$

(**B2**) $\mathbb{J} : X \to X$ is a bounded linear operator satisfying

$$\mathbb{J}^{-1} = -\mathbb{J}, \quad \langle Q_0 \mathbb{J}u, \mathbb{J}u\rangle = \langle Q_0 u, u\rangle, \quad \forall u \in X.$$

Let $J = \mathbb{J} Q_0^{-1} : X^* \supset Q_0(X) \to X$.

(**B3**) $L : X \to X^*$ is a bounded symmetric linear operator such that $L_1 = L - Q_1$ satisfies

$$|\langle L_1 u, v\rangle|^2 \le c_0(\langle Q_0 u, u\rangle\langle Q_0 v, v\rangle + \langle Q_0 u, u\rangle\langle Q_1 v, v\rangle + \langle Q_1 u, u\rangle\langle Q_0 v, v\rangle).$$

(**B4**) There exist closed subspaces $X_\pm \subset X$ such that

$$X = X_- \oplus \ker L \oplus X_+, \quad n^-(L) \triangleq \dim X_- < \infty, \tag{2.27}$$

$$\pm \langle Lu_\pm, u_\pm\rangle > 0, \ \langle Lu_+, u_-\rangle = 0, \ \forall\, 0 \neq u_\pm \in X_\pm. \tag{2.28}$$

(**B5**) Subspaces $X_\pm$ satisfy

$$\ker i_{X_+}^* = \{f \in X^* \mid \langle f, u\rangle = 0, \forall u \in X_+\} \subset Q_0(X) = D(J)$$

where $i_{X_+}^* : X^* \to X_+^*$ is the dual operator of the embedding i_{X_+}.

Obviously the assumption in (**B1**) that $Q_1 + Q_0$ is the Riesz representation of the inner product can be weakened to that it is the Riesz representation of an equivalent inner product. It is also easy to verify that J is closed and anti-symmetric, namely, $J \subset -J^*$. Roughly the L-orthogonal decomposition of X can be constructed a.) by taking $\ker L \oplus X_+$ as the L-orthogonal complement of a carefully chosen X_- and then X_+ as any complimentary subspace of $\ker L$ there; or b.) from a spectral decomposition of the linear operator on X corresponding to the quadratic form $\langle L\cdot, \cdot\rangle$ through certain inner product. In a typical application as in Section 11.6, Q_1 is often a uniformly positive elliptic operator of order $2s$, L_1 is a perturbation containing lower order derivatives with variable coefficients, and Q_0 corresponds to the L^2 duality. It is convenient to start with $X = H^s$ initially. The assumption $n^-(L) < \infty$ may come from the construction of the steady state via some variational approach. The lack of a positive lower bound of L restricted to $X_+ \subset H^s$ is often due to the missing control of the L^2 norm by $\langle L\cdot, \cdot\rangle$. This also forces us to make the slightly stronger assumption (**B5**) than (**H3**). In Chapter 10 we prove

THEOREM 2.7. *There exists a Hilbert space Y such that*
(a) X is densely embedded into Y;
(b) L can be extended to a bounded symmetric linear operator $L_Y : Y \to Y^*$;
*(c) (Y, L_Y, J_Y) satisfy **(H1-3)**, where $J_Y : D(J) \cap Y^* \to Y$ is the restriction of J.*

It is natural to define Y through the completion of X under a norm based on L. To prove this theorem, the key is to show (**H1**) and (**H3**) are satisfied.

2.7. Some Applications to PDEs

We briefly discuss the applications of the general theory to several PDE models in Chapter 11. First, we consider the stability of traveling waves of dispersive wave models of KDV, BBM and good Boussinesq types. These PDE models arise as approximation long wave models for water waves etc. We treat general dispersion symbols including nonlocal ones.

For solitary waves, the linearized equations are written in a Hamiltonian form where the symplectic operators J turn out to be non-invertible unbounded operators. The index formula and the exponential trichotomy estimates are obtained from Theorems 2.2 and 2.3.

For periodic waves, the linearized equations for perturbations of the same period are again written in the Hamiltonian form with J having nontrivial kernels. This brings changes to the index counting formula and stability criteria. In recent years, similar index formula had been studied in various cases. Our results give a unified treatment for general dispersion symbols. For both solitary waves and periodic waves, the linear stability conditions are also shown to imply nonlinear orbital stability. For the unstable cases, the exponential dichotomy can be used to show nonlinear instability and even to further construct local invariant (stable, unstable and center) manifolds near the traveling wave orbit in the energy space. Moreover, when a.) the negative dimension of the linearized energy functional is equal to the unstable dimension of the linearized equation and b.) the kernel of the linearized energy functional is generated exactly by the symmetry group of the system, the orbital stability and local uniqueness on the center manifold could be obtained. These invariant manifolds also give a complete description of dynamics near the orbit of unstable profiles. For more details, we refer to recent papers ([**37**] [**38**]) on the construction of invariant manifolds near unstable traveling waves of supercritical KDV equation and 3D Gross-Pitavaeskii equation.

We then consider the linearized problems arisen from the modulational (Benjamin-Feir, side-band) instability of period waves. Besides obtaining an index formula for each Floquet-Block problem, we also carry out some perturbation analysis to justify that unstable modes in the long wave limit can only arise from zero eigenvalue of the co-periodic problem. Subsequently we obtain the semigroup estimates for both multi-periodic and localized perturbations, which played an important role on the recent proof ([**36**]) of nonlinear modulational instability of various dispersive models.

As another application, we consider the eigenvalue problem of the form $Lu = \lambda u'$, which arises in the stability of traveling waves of generalized Bullough–Dodd equation (11.43). Let $J = \partial_x^{-1}$, then it is equivalent to the Hamiltonian form $JLu = \lambda u$. Thus general theorems can be applied to get instability index formula and the stability criterion which generalize the results in [**69**] by relaxing some restrictions. In particular it implies the linear instability of any traveling wave of generalized Bullough–Dodd equation (11.43), removing the convexity assumption in [**69**].

Next, we consider stability/instability of steady flows of 2D Euler equation in a bounded domain. For a large class of steady flows, the linearized Euler equation can be written in a Hamiltonian form satisfying **(H1)-(H3)**. Here, the symplectic operator J has an infinite dimensional kernel. The index formula is obtained in terms of a reduced operator related to the projection to $\ker L$. By using the perturbation theory in Section 2.5, the structural instability in the case of the presence of embedded eigenvalues is shown. The Hamiltonian structures are also useful in studying the enhanced damping and inviscid damping problems.

Lastly, we study the stability of traveling waves of 2D nonlinear Schrödinger equations with nonzero condition at infinity. When written in the Hamiltonian form JL, the quadratic form $\langle L\cdot, \cdot\rangle$ does not have uniform lower bound on the positive subspace X_+. The strategy used for the 3D case ([**53**]) does not work in 2D. We use the theory in Section 2.6 to construct a new and larger phase space to recover the uniform positivity of $\langle L\cdot, \cdot\rangle$ on the positive space. Then the theory in Section 2.4 is used to prove the stability criterion in terms of the sign of dP/dc, where $P(c)$ is the momentum of a traveling wave of speed c. As a somewhat unusual application of the index formula, we prove the positivity of the momentum P for traveling waves (in both 2D and 3D) with general nonlinear terms.

CHAPTER 3

Basic properties of Linear Hamiltonian systems

In this chapter, we present a few basic qualitative properties of the linear equation (2.1), including the conservation of energy, some elementary spectral properties, *etc.* As our problem is set up in a functional analysis theoretical framework, in some cases we have to follow the painful rigor at an orthodox level. To make it less tedious, we only keep those basic results directly related to the dynamics of (2.1) in this chapter, while some more elementary properties of (2.1), including its well-posedness (Proposition A.1), are left in the Appendix.

Like any Hamiltonian flow, we have the conservation of energy and the symplectic structure of the flow defined by (2.1).

LEMMA 3.1 ([**59**]). *For any solutions $u(t), v(t)$ of (2.1), then we have*

(1) $\frac{d}{dt}\langle Lu(t), v(t)\rangle = 0$;
(2) *$R(J)$ is invariant under e^{tJL}; and*
(3) *if J is one-to-one (J^{-1} not necessarily bounded) and $u(0) \in R(J)$, then* $\frac{d}{dt}\langle J^{-1}u(t), v(t)\rangle = 0$.

PROOF. Property (1) is clearly true if $u(0), v(0) \in D(JL)$ and then the general case follows immediately from a density argument. To prove (2), we first notice, for $x \in D(JL)$,

$$e^{tJL}x - x = J\int_0^t Le^{t'JL}x dt'.$$

Since J is closed and $D(JL)$ is dense, a density argument implies that $\int_0^t Le^{t'JL}x dt' \in D(J)$ for all x and the above equality holds for all x and thus $R(J)$ is invariant under e^{tJL}. For (3), first consider $v(0) \in D(JL)$ and the above equality yields

$$\frac{d}{dt}\langle J^{-1}u(t), v(t)\rangle = \langle Lu(t), v(t)\rangle + \langle J^{-1}u(t), JLv(t)\rangle = 0$$

where we used the assumption that J is anti-self-adjoint. Again the general case of (3) follows from the density of $D(JL)$. □

An immediate consequence of the conservation of the quadratic form $\langle L\cdot,\cdot\rangle$ is on invariant subspaces.

LEMMA 3.2. *Suppose a subspace $X_1 \subset X$ is invariant under e^{tJL}, i.e. $e^{tJL}X_1 \subset X_1$ for all $t \in \mathbf{R}$, then $e^{tJL}X_2 \subset X_2$ for all $t \in \mathbf{R}$ where the closed subspace $X_2 = \{u \in X \mid \langle Lu, v\rangle = 0, \forall v \in X_1\}$.*

PROOF. For any $u \in X_2$, $v \in X_1$, and $t \in \mathbf{R}$, Lemma 3.1 and the invariance of X_1 imply

$$\langle Le^{tJL}u, v\rangle = \langle Lu, e^{-tJL}v\rangle = 0$$

which yields the conclusion. □

While in a substantial part of the paper, we shall work with the real Hilbert space X and real operators J, L, *etc.*, for considerations where complex eigenvalues are involved, we have to work with their standard **complexification**. See the Appendix for details.

Let λ be an eigenvalue of JL (i.e. $\lambda \in \sigma(JL)$) and

$$E_\lambda = \{u \in X \mid (JL - \lambda I)^k u = 0, \text{ for some integer } k \geq 1\}.$$

Then by Lemma 3.1, we have

LEMMA 3.3 (Lemma 2 in [**59**] or Lemma 2.7 in [**32**]). *If $v_1 \in E_{\lambda_1}, v_2 \in E_{\lambda_2}$ and $\lambda_1 + \bar{\lambda}_2 \neq 0$, then $\langle Lv_1, v_2 \rangle = 0$.*

The following lemma will be repeatedly used to analyze the structure of $E_{i\mu}$, $\mu \in \mathbf{R}$.

LEMMA 3.4. *For any $i\mu \in \sigma(JL)$ $(\mu \in \mathbf{R})$, $u_1 \in (JL - i\mu)^l X$, and $u_2 \in \ker(JL - i\mu)^l$, then $\langle Lu_1, u_2 \rangle = 0$.*

PROOF. First, we observe that for any $u, v \in X$,

$$\langle L(JL - i\mu)u, v \rangle = -\langle Lu, (JL - i\mu)v \rangle. \tag{3.1}$$

Let $v \in X$ such that $(JL - i\mu)^l v = u_1$, then we have

$$\langle Lu_1, u_2 \rangle = \langle L(JL - i\mu)^l v, u_2 \rangle = (-1)^l \langle Lv, (JL - i\mu)^l u_2 \rangle = 0.$$

□

The following lemma is a direct consequence of Lemma 3.4 and (3.1).

LEMMA 3.5. *For any $i\mu \in \sigma(JL) \cap i\mathbf{R}$, it holds*

$$E_{i\mu} = \ker(JL - i\mu)^{2k^{\leq 0}(i\mu)+1}, \ \mu \neq 0, \quad \textit{and} \quad E_0 = \ker(JL)^{2k_0^{\leq 0}+2}.$$

REMARK 3.1. *As $JL - i\mu$ is a generator of a strongly C^0 semigroup, $(JL - i\mu)^m$ is closed for any m and thus $E_{i\mu}$ is a closed subspace and $JL|_{E_{i\mu}}$ is a bounded operator with $\sigma(JL|_{E_{i\mu}}) = \{i\mu\}$.*

PROOF. We first consider $\mu \neq 0$ and argue by contradiction. Suppose $u \in E_{i\mu}$ such that

$$(JL - i\mu)^K u = 0, \quad (JL - i\mu)^{K-1} u \neq 0, \quad K \geq 2k^{\leq 0}(i\mu) + 2.$$

For any $K - 1 \geq j_1, j_2 \geq K - k^{\leq 0}(i\mu) - 1$, we obtain from Lemma 3.4

$$\langle L(JL - i\mu)^{j_1} u, (JL - i\mu)^{j_2} u \rangle = 0.$$

Therefore the quadratic form $\langle L\cdot, \cdot \rangle$ vanishes on

$$span\{(JL - i\mu)^{K-1} u, (JL - i\mu)^{K-2}, \ldots, (JL - i\mu)^{K - k^{\leq 0}(i\mu) - 1} u\},$$

whose dimension is $k^{\leq 0}(i\mu) + 1$. This contradicts the definition of $k^{\leq 0}(i\mu)$.

To finish the proof, we consider $\mu = 0$. Again we argue by contradiction. Suppose $u \in E_0$ is such that

$$(JL)^K u = 0, \quad (JL)^{K-1} u \neq 0, \quad K \geq 2k_0^{\leq 0} + 3.$$

Case 1. $(JL)^{K-1} u \notin \ker L$. In this case, clearly

$$span\{(JL)^j u \mid 0 \leq j \leq K - 1\} \cap \ker L = \{0\}.$$

Let $\tilde{E}_0 \subset E_0$ be a subspace such that $E_0 = \tilde{E}_0 \oplus \ker L$ and $(JL)^j u \in \tilde{E}_0$ for any $0 \leq j \leq K-1$. Much as in the above, $\langle L\cdot, \cdot\rangle$ vanishes on

$$Z \triangleq span\{(JL)^{K-1}u, (JL)^{K-2}u, \ldots, (JL)^{K-k_0^{\leq 0}-1}u\} \subset \tilde{E}_0.$$

Since $\dim Z = k_0^{\leq 0} + 1$, this is a contradiction to the definition of $k_0^{\leq 0}$.

Cases 2. $(JL)^{K-1}u \in \ker L \backslash \{0\}$. Clearly

$$span\{(JL)^j u \mid 0 \leq j \leq K-2\} \cap \ker L = \{0\}.$$

Let $\tilde{E}_0 \subset E_0$ be a subspace such that $E_0 = \tilde{E}_0 \oplus \ker L$ and $(JL)^j u \in \tilde{E}_0$ for any $0 \leq j \leq K-2$. Let

$$Z \triangleq span\{(JL)^{K-2}u, (JL)^{K-3}u, \ldots, (JL)^{K-k_0^{\leq 0}-2}u\} \subset \tilde{E}_0.$$

According to Lemma 3.4, for $K - k_0^{\leq 0} - 2 \leq j_1, j_2 \leq K-2$ and $j_1 + j_2 \geq K$, we have $\langle L(JL)^{j_1}u, (JL)^{j_2}u\rangle = 0$. If $K - k_0^{\leq 0} - 2 \leq j_1, j_2 \leq K-2$ and $j_1 + j_2 < K$, it must hold $j_1 = j_2 = K - k_0^{\leq 0} - 2$ and $K = 2k_0^{\leq 0} + 3$. Using (3.1) we obtain

$$\langle L(JL)^{K-k_0^{\leq 0}-2}u, (JL)^{K-k_0^{\leq 0}-2}u\rangle = (-1)^{K-k_0^{\leq 0}-2}\langle L(JL)^{K-1}u, u\rangle = 0$$

where in the last equality we used $(JL)^{K-1}u \in \ker L$. Therefore $\langle L\cdot, \cdot\rangle$ vanishes on Z. Since $\dim Z = k_0^{\leq 0} + 1$, this is again a contradiction to the definition of $k_0^{\leq 0}$. The proof of the lemma is complete. □

To end the chapter of basic properties, we prove the following Lemma on the symmetry of $\sigma(JL)$ about both axes.

LEMMA 3.6. *Assume **(H1)**-**(H3)**, except for $n^-(L) < \infty$. Suppose $\lambda \in \sigma(JL)$, then we have*

i) $\pm\lambda, \pm\bar{\lambda} \in \sigma(JL)$.

ii) Suppose λ is an eigenvalue of JL and assume in addition $\ker L = \{0\}$ or $\lambda \neq 0$, then $\bar{\lambda}$ is also an eigenvalue of JL and $-\lambda, -\bar{\lambda}$ are eigenvalues of $(JL)^* = -LJ$. Moreover for any $k > 0$,

$$\ker(JL - \bar{\lambda})^k = \{\bar{u} \mid u \in \ker(JL - \lambda)^k\} \tag{3.2}$$

and

$$L : \ker(JL - \lambda)^k \to \ker\left((JL)^* + \bar{\lambda}\right)^k = \ker(LJ - \bar{\lambda})^k \tag{3.3}$$

is an anti-linear isomorphism.

iii) Suppose λ is an isolated eigenvalue of JL with finite algebraic multiplicity, then $-\lambda, \pm\bar{\lambda}$ are also eigenvalues of JL with the same algebraic and geometric multiplicities.

Here the operators J and L are understood as their complexification, and thus are anti-linear mappings satisfying (A.9).

PROOF. As i) is trivial if $\lambda = 0$, so we assume $\lambda \neq 0$ or $\ker L = \{0\}$.

Due to (A.12) which states that JL is real, (3.2) and $\bar{\lambda} \in \sigma(JL)$ follow immediately. We are left to prove $-\lambda, -\bar{\lambda} \in \sigma(JL)$ and (3.3).

The anti-linearity property (A.9) implies

$$L(JL - \lambda)u = (LJ - \bar{\lambda})Lu = -((JL)^* + \bar{\lambda})Lu, \ \forall u \in D(JL). \tag{3.4}$$

Therefore we have that, for any integer $k > 0$,

$$\left((JL)^* + \bar{\lambda}\right)^k Lu = (-1)^k L(JL - \lambda)^k u, \ \forall u \in D\left((JL)^k\right). \tag{3.5}$$

It follows from (3.5) that $L\big(\ker(JL-\lambda)^k\big) \subset \ker\big((JL)^* + \bar{\lambda}\big)^k$. Under the assumption $\lambda \neq 0$ or $\ker L = \{0\}$, it holds $\ker L \cap E_\lambda = \{0\}$ and thus L is one-to-one on E_λ. Therefore if λ is an eigenvalue of JL, then E_λ is nontrivial which implies $L\big(\ker(JL-\lambda)^k\big)$, as well as $\ker\big((JL)^* + \bar{\lambda}\big)^k$, are nontrivial. We obtain that $-\bar{\lambda}$, as well as $-\lambda$, is an eigenvalue of $(JL)^*$. Consequently $-\lambda, -\bar{\lambda} \in \sigma(JL)$.

To finish the proof of (3.3), we only need to show

$$L\big(\ker(JL-\lambda)^k\big) \supset \ker\big((JL)^* + \bar{\lambda}\big)^k.$$

This is obvious from (3.5) if $\ker L = \{0\}$. In the case of $\lambda \neq 0$, it is clear $\ker\big((JL)^* + \bar{\lambda}\big)^k \subset R(L)$. Therefore for any $v \in \ker\big((JL)^* + \bar{\lambda}\big)^k$, there exists $u_1 \in X$ such that $v = Lu_1$. Equation (3.5) again implies

$$w = (JL-\lambda)^k u_1 \in \ker L.$$

As $\lambda \neq 0$, let $u = u_1 - (-\lambda)^{-k} w$, then since $(JL-\lambda)\, w = (-\lambda)\, w$, we have $v = Lu$ and $u \in \ker(JL-\lambda)^k$ due to $(JL-\lambda)^k w = (-\lambda)^k\, w$. Therefore $\ker\big((JL)^* + \bar{\lambda}\big)^k \subset L\ker(JL-\lambda)^k$ and thus L is an one to one correspondence (actually an anti-linear isomorphism) from $\ker(JL-\lambda)^k$ to $\ker\big((JL)^* + \bar{\lambda}\big)^k$.

Finally, suppose $\lambda \neq 0$ and $R(JL-\lambda) \neq X$, we will show $R\big((JL)^* + \bar{\lambda}\big) \neq X^*$ which implies $-\bar{\lambda}, -\lambda \in \sigma\big((JL)^*\big) = \overline{\sigma(JL)}$ and thus completes the proof of (i). Assume one the contrary $R\big((JL)^* + \bar{\lambda}\big) = X^*$. Let $\gamma \in D(J)\backslash R(L)$, according to Remark 2.4, there exists $u \in \ker L$ such that $\langle \gamma, u\rangle \neq 0$. One can compute $\langle \big((JL)^* + \bar{\lambda}\big)\gamma, u\rangle = \bar{\lambda}\langle \gamma, u\rangle \neq 0$ and thus $\big((JL)^* + \bar{\lambda}\big)\gamma \notin R(L)$. Therefore, if $R\big((JL)^* + \bar{\lambda}\big) = X^*$, it must hold $\big((JL)^* + \bar{\lambda}\big)\big(R(L)\big) = R(L)$, which is the range of the right side of (3.4). However, since $\lambda \neq 0$, we have $(JL-\lambda)(\ker L) = \ker L$. Along with $R(JL-\lambda) \neq X$, it implies $R(L) \not\subset R\big(L(JL-\lambda)\big)$, which is the range of the left side of (3.4). We obtain a contradiction and thus $R\big((JL)^* + \bar{\lambda}\big) \neq X^*$.

If $\lambda \in \sigma(JL)$ is isolated and of finite multiplicity, then the same is true for $\bar{\lambda}$. By i) and ii), $-\lambda, -\bar{\lambda} \in \sigma\big((JL)^*\big)$ are also isolated and of the same multiplicities, this implies that $-\lambda, -\bar{\lambda} \in \sigma(JL)$ have the same (geometric and algebraic) multiplicities (see [**42**] p. 184). □

REMARK 3.2. *As in the proof of Lemma* A.3 *and Corollary* A.3, *the assumption* $n^-(L) < \infty$ *is not required in the above proof. So Lemma* 3.6 *holds even when* $n^-(L) = \infty$. *On the other hand, this lemma gives the symmetry of* $\sigma(JL)$, *but not for general eigenvalues, except for purely imaginary eigenvalues or isolated eigenvalues of finite multiplicity. If* $\lambda \in \sigma(JL)$ *is a nonzero eigenvalue which is non-isolated or of infinite multiplicity, then above lemma implies that* $-\lambda, -\bar{\lambda}$ *are eigenvalues of* $\sigma\big((JL)^*\big)$. *In general, we can not exclude the possibility that* $-\lambda, -\bar{\lambda}$ *are not eigenvalues of* JL. *However, when* $n^-(L) < \infty$, *any* $\lambda \in \sigma(JL)$ *with* $\mathrm{Re}\,\lambda \neq 0$ *must be isolated and of finite multiplicity, and the symmetry of eigenvalues and the dimensions of their eigenspaces are given in Corollary* 6.1.

CHAPTER 4

Finite dimensional Hamiltonian systems

In this chapter, we consider the case where the energy space X of (2.1) is $X = \mathbf{R}^n$ which is complexified to $\mathbf{C}^n$. The assumptions (**H.1-3**) become that J is a real anti-symmetric $n \times n$ matrix and L is a real symmetric $n \times n$ matrix. The counting formula (2.13) essentially follows from [**59**], except for the formula (2.16). We do not need to assume that J is invertible as assumed in [**59**].

For $\lambda \in \sigma(JL)$, define

$$I_\lambda = E_\lambda \oplus E_{-\bar{\lambda}} \text{ if } \lambda \notin i\mathbf{R}, \text{ and } I_\lambda = E_\lambda \text{ if } \lambda \in i\mathbf{R}.$$

We have $\mathbf{C}^n = I_{\lambda_1} \oplus \cdots \oplus I_{\lambda_l}$, where $\lambda_j \in \sigma(JL)$ are all distinct eigenvalues of JL with $\mathrm{Re}\lambda_j \geq 0$. By Lemma 3.3, we have

$$n^-(L) = \sum_j n^-\left(L|_{I_{\lambda_j}}\right). \tag{4.1}$$

Based on Lemma 3.6 of the symmetry of $\sigma(JL)$, to prove Theorem 2.3 in the finite dimensional case, it suffices to compute $n^-(L|_{I_\lambda})$ for any $0 \neq \lambda \in \sigma(JL) \backslash i\mathbf{R}$.

LEMMA 4.1 ([**59**]). *Let $\lambda \in \sigma(JL)$. Assume $\ker L = \{0\}$ or $\lambda \neq 0$, then the restriction $\langle L\cdot, \cdot\rangle|_{I_\lambda}$ is non-degenerate.*

PROOF. Suppose $\langle L\cdot, \cdot\rangle|_{I_\lambda}$ is degenerate. Then there exists $0 \neq u \in I_\lambda$ such that $\langle Lu, v\rangle = 0$ for any $v \in I_\lambda$. Since $\mathbf{C}^n$ is the direct sum of all different $I_{\lambda'}$, $\lambda' \in \sigma(JL)$, this implies that $\langle Lu, v\rangle = 0$ for any $v \in \mathbf{C}^n$ by Lemma 3.3. So $Lu = 0$ and thus $0 \neq u \in I_\lambda \cap \ker L$. It implies that $\lambda = 0$ and $\ker L \neq \{0\}$, a contradiction to our assumptions. □

LEMMA 4.2 ([**59**] or [**32**]). *If $\mathrm{Re}\,\lambda > 0$ and let m_λ to be the algebraic multiplicity of λ. Then $n^-(L|_{I_\lambda}) = m_\lambda$.*

PROOF. From Lemma 3.3, the quadratic form $\langle L\cdot, \cdot\rangle$ on $I_\lambda = E_\lambda \oplus E_{-\bar{\lambda}}$ can be represented in the block form $\begin{pmatrix} 0 & A \\ A^* & 0 \end{pmatrix}$. Lemma 4.1 implies the non-degeneracy of A and thus the lemma follows. □

The counting formula (2.13) in the finite dimensional case follows from these lemmas and (4.1).

In the rest of this section, we carefully analyze $k^-(i\mu) = n^-\left(L|_{E_{i\mu}}\right)$, $\mu \in \mathbf{R}$, and obtain Proposition 2.2 in finite dimensions. Based on Lemma 3.4 and equation (3.1), we first prove

LEMMA 4.3. *Suppose $i\mu \in \sigma(JL)$ ($\mu \in \mathbf{R}$) and $K > 0$ is an integer, then*

(1) *for $u, v \in \ker(JL - i\mu)^K$,*

$$Q_K(u, v) \triangleq i^{K-1}\langle L(JL - i\mu)^{K-1}u, v\rangle$$

defines a Hermitian form on $\ker(JL - i\mu)^K$*; and*

(2) *assume* $\ker L = \{0\}$ *or* $\mu \neq 0$*, then*

$$Y_K \triangleq \big(\ker(JL - i\mu)^K \cap R(JL - i\mu)\big) + \ker(JL - i\mu)^{K-1} = \ker Q_K.$$

PROOF. That Q_K is a Hermitian form on $\ker(JL - i\mu)^K$ is an immediate consequence of equation (3.1). Lemma 3.4 also implies $Y_K \subset \ker Q_K$. We will show $Y_K = \ker Q_K$ under the additional assumption $\ker L = \{0\}$ or $\mu \neq 0$. Suppose $u \in \ker(JL - i\mu)^K$ is such that

$$Q_K(u, v) = i^{K-1}\langle L(JL - i\mu)^{K-1}u, v\rangle = 0, \quad \forall\, v \in \ker(JL - i\mu)^K.$$

By duality, it implies that

$$L(JL - i\mu)^{K-1}u \in \big((JL - i\mu)^K)\big)^*(\mathbf{C}^n) = (LJ - i\mu)^K(\mathbf{C}^n).$$

Therefore there exists $w \in \mathbf{C}^n$ such that

$$L(JL - i\mu)^{K-1}u = (LJ - i\mu)^K w.$$

Since $\mu \neq 0$ or L is surjective, the above equation implies $w \in R(L)$ and thus there exists $\tilde{w} \in \mathbf{C}^n$ such that $w = L\tilde{w}$. Consequently,

$$(LJ - i\mu)^{K-1}L\big(u - (JL - i\mu)\tilde{w}\big) = L(JL - i\mu)^{K-1}u - (LJ - i\mu)^K w = 0$$

which along with Lemma 3.6 implies

$$L\big(u - (JL - i\mu)\tilde{w}\big) \in \ker(LJ - i\mu)^{K-1} = L\ker(JL - i\mu)^{K-1}.$$

Therefore there exists $v \in \ker(JL - i\mu)^{K-1}$ such that

$$y = u - (JL - i\mu)\tilde{w} - v \in \ker L.$$

If $\mu \neq 0$, let $w_1 = \tilde{w} + \frac{1}{i\mu}y$. If $\ker L = \{0\}$, we have $y = 0$ and let $w_1 = \tilde{w}$. In both cases, we have

$$u = v + (JL - i\mu)w_1, \quad v \in \ker(JL - i\mu)^{K-1} \subset \ker(JL - i\mu)^K.$$

Therefore $(JL - i\mu)w_1 \in R(JL - i\mu) \cap \ker(JL - i\mu)^K$ and then $u \in Y_K$. The proof is complete. □

COROLLARY 4.1. *Assume* $\ker L = \{0\}$ *or* $\mu \neq 0$*, then* Q_K *induces a non-degenerate Hermitian form on the quotient space* $\ker(JL - i\mu)^K/Y_K$.

If K is odd, for any $u, v \in \ker(JL - i\mu)^K$, clearly

$$Q_K(u, v) = \langle L(JL - i\mu)^{\frac{K-1}{2}}u, (JL - i\mu)^{\frac{K-1}{2}}v\rangle. \tag{4.2}$$

DEFINITION 4.1. *For odd* K*, define* $n_K^-(i\mu)$ *to be the negative index of the quadratic form* Q_K.

The above quotient space $\ker(JL - i\mu)^K/Y_K$ is closely related to Jordan chains. Suppose a basis of $\mathbf{C}^n$ realizes the Jordan canonical form of JL, and there are totally l Jordan blocks of size $K \times K$ corresponding to $i\mu$. There must be l Jordan chains of length K in such basis, each of which is generated by some $v \in \ker(JL - i\mu)^K/Y_K$ as

$$v,\ (JL - i\mu)v,\ \ldots,\ (JL - i\mu)^{K-1}v.$$

From standard linear algebra, we have the following lemma.

LEMMA 4.4. *Vectors* $v_{1,1}, \dots, v_{l,1}$ *generate all* l *Jordan chains of length* K *in the sense that*

$$v_{j,k} = (JL - i\mu)^{k-1} v_{j,1}, \quad 1 \le k \le K, \ 1 \le j \le l,$$

are in a basis of $\mathbf{C}^n$ *realizing all* l *Jordan blocks of size* K *of* JL *corresponding to* $i\mu \in \sigma(JL)$, *if and only if*

$$v_{1,1} + Y_K, \ \dots, \ v_{l,1} + Y_K$$

form a basis of $\ker(JL - i\mu)^K / Y_K$.

The following lemma would lead to the realization of the Jordan canonical form of JL and skew-diagonalization of L simultaneously.

LEMMA 4.5. *Assume* $\ker L = \{0\}$ *or* $\mu \ne 0$ *where* $i\mu \in \sigma(JL) \cap i\mathbf{R}$. *Suppose* $\dim \ker(JL - i\mu)^K / Y_K = l > 0$ *and* $Z \subset \ker(JL - i\mu)^K$ *satisfies*

$$JL(Z) = Z \quad \text{and} \quad Z/(Y_K \cap Z) = \ker(JL - i\mu)^K / Y_K,$$

then there exist $v_1, \dots, v_l \in Z$ *such that*

$$\langle L(JL - i\mu)^m v_j, v_k \rangle = \pm i^{K-1} \delta_{j,k} \delta_{m,K-1}, \quad 0 \le m \le K-1. \tag{4.3}$$

PROOF. Since Q_K induces a non-degenerate Hermitian form on $\ker(JL - i\mu)^K / Y_K = Z/(Z \cap Y_K)$, there exist $w_1, \dots, w_l \in Z$ such that $w_1 + Y_K, \dots, w_l + Y_K$ form a basis of $\ker(JL - i\mu)^K / Y_K$ and diagonalize Q_K, that is,

$$\langle L(JL - i\mu)^{K-1} w_j, w_k \rangle = (-i)^{K-1} Q_K(w_j, w_k) = \pm i^{K-1} \delta_{j,k}.$$

Therefore we have found $w_1, \dots, w_l$ satisfying (4.3) for $m = K - 1$.

Suppose $1 \le m_0 + 1 \le K - 1$ and we have found $w_1, \dots, w_l \in Z$ satisfying (4.3) for $m \ge m_0 + 1$. Denote

$$\alpha = K - 1 - m_0 \ge 1, \ Q_K(w_j, w_k) = b_{j,k} = \pm \delta_{j,k}, \ \langle L(JL - i\mu)^{m_0} w_j, w_k \rangle = c_{j,k}.$$

In the next step we will construct $v_1, \dots, v_l$ satisfying (4.3) for $m \ge m_0$ in the form of

$$v_j = w_j + \sum_{j'=1}^{j} a_{j,j'} (JL - i\mu)^\alpha w_{j'} \in Z.$$

According to (3.1), $\langle L(JL - i\mu)^m \cdot, \cdot \rangle$ is Hermitian or anti-Hermitian. Without loss of generality, we may consider only $j \le k$ in (4.3). Compute using (3.1)

$$\begin{aligned}
\langle L(JL - i\mu)^m v_j, v_k \rangle = {} & (-1)^\alpha \sum_{k'=1}^{k} \overline{a_{k,k'}} \langle L(JL - i\mu)^{\alpha+m} w_j, w_{k'} \rangle \\
& + \langle L(JL - i\mu)^m w_j, w_k \rangle + \sum_{j'=1}^{j} a_{j,j'} \langle L(JL - i\mu)^{\alpha+m} w_{j'}, w_k \rangle \\
& + (-1)^\alpha \sum_{j'=1}^{j} \sum_{k'=1}^{k} a_{j,j'} \overline{a_{k,k'}} \langle L(JL - i\mu)^{2\alpha+m} w_{j'}, w_{k'} \rangle.
\end{aligned} \tag{4.4}$$

If $m + \alpha = K - 1 - m_0 + m \ge K$, the induction assumption and the above equation imply

$$\langle L(JL - i\mu)^m v_j, v_k \rangle = \langle L(JL - i\mu)^m w_j, w_k \rangle = \pm i^{K-1} \delta_{j,k} \delta_{m,K-1}$$

and thus (4.3) for $m \geq m_0 + 1$ holds for these $v_1, \ldots, v_l$ with any choices of $a_{j,j'}$. For $m = m_0$, i.e. $m + \alpha = K - 1$, if $j < k$, (4.4) implies

$$\langle L(JL - i\mu)^{m_0} v_j, v_k \rangle = c_{j,k} + (-1)^\alpha (-i)^{K-1} \overline{a_{k,j}} b_{j,j}.$$

Noticing $b_{j,j} = \pm 1$ and letting

$$a_{k,j} = (-1)^{\alpha+1} (-i)^{K-1} \overline{c_{j,k}} b_{j,j}, \ j < k,$$

then we have

$$\langle L(JL - i\mu)^{m_0} v_j, v_k \rangle = 0, \quad j < k.$$

If $j = k$,

$$\langle L(JL - i\mu)^{m_0} v_k, v_k \rangle = c_{k,k} + (-i)^{K-1} b_{k,k} \big(a_{k,k} + (-1)^\alpha \overline{a_{k,k}} \big).$$

Let

$$a_{k,k} = -\frac{1}{2} i^{K-1} b_{k,k} c_{k,k} = -\frac{1}{2} i^\alpha b_{k,k} i^{m_0} c_{k,k}.$$

Since (3.1) implies $c_{k,j} = (-1)^{m_0} \overline{c_{j,k}}$, we have $i^{m_0} c_{k,k} \in \mathbf{R}$ and thus $i^\alpha a_{k,k} \in \mathbf{R}$ which makes it easy to verify

$$\langle L(JL - i\mu)^{m_0} v_k, v_k \rangle = 0.$$

Therefore $v_1, \ldots, v_l \in Z$ satisfy (4.3) for all $m \geq m_0$ and the lemma follows from the induction. □

We are in a position to prove Proposition 2.2 in finite dimensions.

Proof of Proposition 2.2 assuming $\dim X < \infty$ **and** $\ker L = \{0\}$: Let $E^D = \{0\}$, then $1 < k_1 < \cdots < k_{j_0}$ are the dimensions of nontrivial Jordan blocks in $E_{i\mu}$, $\mu \in \mathbf{R}$, and there are $l_j > 0$ Jordan blocks of size k_j. For each $1 \leq j \leq j_0$, we will find linearly independent

$$\{ u_{p,q}^{(j)} \mid p = 1, \ldots, l_j, \ q = 1, \ldots, k_j \} \subset E_{i\mu}$$

which form all Jordan chains of length k_j and satisfy the desired properties. The construction is by induction on j.

For $j = j_0$, applying Lemma 4.5 to $Z = \ker(JL - i\mu)^{k_{j_0}} = E_{i\mu}$, where $\dim \ker(JL - i\mu)^{k_{j_0}} / Y_{k_{j_0}} = l_{j_0}$ according to Lemma 4.4, there exist $u_{1,1}^{(j_0)}, \ldots, u_{l_{j_0},1}^{(j_0)}$ such that

$$\langle L(JL - i\mu)^m u_{p_1,1}^{(j_0)}, u_{p_2,1}^{(j_0)} \rangle = \pm i^{k_{j_0}-1} \delta_{p_1,p_2} \delta_{m,k_{j_0}-1}, \quad 0 \leq m \leq j_0 - 1. \tag{4.5}$$

In particular we have $Q_K(u_{p_1,1}^{(j_0)}, u_{p_2,1}^{(j_0)}) = \pm \delta_{p_1,p_2}$. Lemma 4.3 and Corollary 4.1 imply that $u_{1,1}^{(j_0)} + Y_{k_{j_0}}, \ldots, u_{l_{j_0},1}^{(j_0)} + Y_{k_{j_0}}$ form a basis of $\ker(JL - i\mu)^{k_{j_0}} / Y_{k_{j_0}}$. From Lemma 4.4, we obtain that

$$u_{p,q}^{(j_0)} = (JL - i\mu)^{q-1} u_{p,1}^{(j_0)}, \quad q = 1, \cdots, k_{j_0}, \ p = 1, \ldots, l_{j_0}$$

form l_{j_0} Jordan chains realizing all Jordan blocks of size k_{j_0} of JL corresponding to $i\mu \in \sigma(JL)$. Moreover, equation (4.5) implies

$$\langle L u_{p_1,q_1}^{(j_0)}, u_{p_2,q_2}^{(j_0)} \rangle = \pm i^{k_{j_0}-1} \delta_{p_1,p_2} \delta_{q_1+q_2,k_{j_0}+1}.$$

Suppose $0 \leq j_* < j_0$ and we have constructed linearly independent $u_{p,q}^{(j)}$ for all $j_* < j \leq j_0$, $1 \leq p \leq l_j$, $1 \leq q \leq k_j$ satisfying

$$u_{p,q}^{(j)} = (JL - i\mu)^{q-1} u_{p,1}^{(j)}, \quad \langle L u_{p_1,q_1}^{(j)}, u_{p_2,q_2}^{(j)} \rangle = \pm i^{k_j-1} \delta_{p_1,p_2} \delta_{q_1+q_2,k_j+1}. \tag{4.6}$$

Clearly

$$Z_1 = span\{u_{p,q}^{(j)} \mid j_* < j \le j_0,\ 1 \le p \le l_j,\ 1 \le q \le k_j\} \subset E_{i\mu}$$

is a subspace invariant under JL. Moreover, vectors $\{u_{p,q}^{(j)}\}$ form a basis of Z_1 realizing the Jordan canonical form of JL on Z_1 consisting of all those Jordan blocks of JL corresponding to $i\mu$ of size greater than k_{j_*}. According to (4.6), the quadratic form $\langle L\cdot,\cdot\rangle$ is non-degenerate on Z_1. In the next step we will construct $u_{p,q}^{(j_*)}$ for $1 \le p \le l_{j_*}$ and $1 \le q \le k_{j_*}$. Let

$$Z = \{u \in E_{i\mu} \mid \langle Lu, v\rangle = 0,\ \forall v \in Z_1\}.$$

Due to the non-degeneracy of $\langle L\cdot,\cdot\rangle$ on both Z_1 and $I_{i\mu} = E_{i\mu}$ (Lemma 4.1), we have $E_{i\mu} = Z_1 \oplus Z$. For any $u \in Z$ and $v \in Z_1$, due to the symmetry of L and J, we have

$$\langle LJLu, v\rangle = -\langle Lu, JLv\rangle = 0, \quad \text{as } JLv \in Z_1$$

which implies $JL(Z) \subset Z$. Since the Jordan canonical form of JL on Z_1 includes all Jordan blocks of JL on $E_{i\mu}$ of size greater than k_{j_*}, the Jordan canonical form of JL on Z must be those Jordan blocks of JL on $E_{i\mu}$ of size no greater than k_{j_*}. Therefore $Z \subset \ker(JL - i\mu)^{k_{j_*}}$ and then Lemma 4.4 implies $Z/(Z \cap Y_{k_{j_*}}) = \ker(JL - i\mu)^{k_{j_*}}/Y_{k_{j_*}}$. Lemma 4.5 provides vectors $u_{1,1}^{(j_*)}, \ldots, u_{l_{j_*},1}^{(j_*)} \in Z$. It is easy to verify that $u_{p,q}^{(j)}$, $j_* \le j \le j_0$, satisfy the induction assumption for $j_* \le j \le j_0$. Therefore, by induction, we find all $u_{p,q}^{(j)}$ satisfying (4.6) and realizing all Jordan blocks of JL on $E_{i\mu}$ of size greater than 1. It is straightforward to verify all the properties in Proposition 2.2. In particular, Lemma 4.4 and equation (4.2) imply that the Krein signature defined in Proposition 2.2 and Remark 2.12 coincides with the one in the above Definition 4.1 in terms of Q_K. Therefore it is independent of the choice of the basis (Jordan chains) realizing the Jordan canonical form.

Finally, let

$$E^1 = \{v \in E_{i\mu} \mid \langle Lu_{p,q}^{(j)}, v\rangle = 0,\ \forall\ 1 \le j \le j_0,\ 1 \le p \le l_j,\ 1 \le q \le k_j\}.$$

Much as in the invariance of Z in the above, $JL(E^1) \subset E^1$. Since all the Jordan blocks are realized by

$$\left\{u_{p,q}^{(j)},\ 1 \le j \le j_0,\ 1 \le p \le l_j,\ 1 \le q \le k_j\right\},$$

we have $E^1 \subset \ker(JL - i\mu)$. This completes the proof. □

Based on Proposition 2.2, we give the following result to be used later.

LEMMA 4.6. *Let J, L be real $n \times n$ matrices. Assume J is anti-symmetric and L is symmetric and nonsingular. Then there exists an invariant (under JL) subspace W of $\mathbf{C}^n$ such that $\dim W = n^-(L)$ and $\langle L\cdot,\cdot\rangle|_W \le 0$.*

PROOF. For any purely imaginary eigenvalue $\lambda = i\mu \in i\mathbf{R}$, we start with the special basis of $E_{i\mu}$ given by Proposition 2.2 (as well as Remark 2.11). For each Jordan chain $\left\{u_{p,1}^{(j)}, \cdots, u_{p,k_j}^{(j)}\right\}$ of even length, define the subspace $Z_{i\mu,j,p} = span\left\{u_{p,1}^{(j)}, \cdots, u_{p,k_j/2}^{(j)}\right\}$. For each Jordan chain $\left\{u_{p,1}^{(j)}, \cdots, u_{p,k_j}^{(j)}\right\}$ of odd length $k_j \ge 1$, define the subspace

$$Z_{i\mu,j,p} = \begin{cases} span\{u_{p,1}^{(j)}, \cdots, u_{p,(k_j-1)/2}^{(j)}\} & \text{if } \langle Lu_{p,(k_j+1)/2}^{(j)}, u_{p,(k_j+1)/2}^{(j)}\rangle > 0, \\ span\{u_{p,1}^{(j)}, \cdots, u_{p,(k_j+1)/2}^{(j)}\} & \text{if } \langle Lu_{p,(k_j+1)/2}^{(j)}, u_{p,(k_j+1)/2}^{(j)}\rangle < 0. \end{cases}$$

Proposition 2.2 implies that $\langle Lu, u\rangle \leq 0$ for all $u \in Z_{i\mu,j,p}$ defined above. For any eigenvalue λ of JL with $\operatorname{Re}\lambda > 0$, recall $\langle Lu, u\rangle = 0$ for all $u \in E_\lambda$ by Lemma 3.3. Define

$$Z_{i\mu} = \oplus_{j=0}^{k_j} \oplus_{p=1}^{l_j} Z_{i\mu,j,p}$$

and

$$W = \oplus_{\operatorname{Re}\lambda>0} E_\lambda \oplus_{i\mu\in\sigma(JL)\cap i\mathbf{R}} Z_{i\mu}.$$

Then $\langle L\cdot,\cdot\rangle|_W \leq 0$ since these subspaces are pairwise orthogonal in $\langle L\cdot,\cdot\rangle$. Moreover, $\dim W = n^-(L)$ due to the counting formula (2.13) and (2.16). $\square$

CHAPTER 5

Invariant subspaces

In this chapter, we study subspaces of X invariant under JL, including both positive and negative results. As the first step to prove our main results, a non-positive (with respect to $\langle L\cdot,\cdot\rangle$) invariant subspace of the maximal possible dimension $n^-(L)$ is derived in Section 5.1. The existence of such subspaces is not only useful for the linear dynamics, but also a rather interesting and delicate result as demonstrated in the discussions and examples in Section 5.2. Throughout this chapter, we work under the non-degeneracy assumption that (2.4) holds for L which is equivalent to $L : X \to X^*$ is an isomorphism.

5.1. Maximal non-positive invariant subspaces (Pontryagin invariant subspaces)

THEOREM 5.1. *In additional to hypotheses* ***(H-3)***, *assume L satisfies the non-degeneracy assumption* (2.4), *then*

(1) $\dim W \leq n^-(L)$ *holds for any subspace* $W \subset X$ *satisfying* $\langle Lu, u\rangle \leq 0$ *for any* $u \in W$; *and*
(2) *there exists a subspace* $W \subset D(JL)$ *such that*

$$\dim W = n^-(L), \quad JL(W) \subset W, \text{ and } \langle Lu, u\rangle \leq 0, \ \forall u \in W.$$

REMARK 5.1. *Though the theorem is stated for real Hilbert spaces, the same proof shows that it also holds for complex Hilbert space X and Hermitian forms L and J. Furthermore, the invariance of W under JL implies that* $W \subset \cap_{k=1}^{\infty} D\big((JL)^k\big)$.

This theorem is basically equivalent to the classical Pontryagin invariant subspace theorem which is usually stated for a self-adjoint operator A with resect to some indefinite quadratic form $\langle L\cdot,\cdot\rangle$ on X with finitely many negative directions (i.e. L satisfies **(H2)** with $n^-(L) < \infty$ and $\ker L = \{0\}$). It states that there exists a subspace $W \subset X$ such that W is invariant under A, $\langle L\cdot,\cdot\rangle|_W \leq 0$ and $\dim W = n^-(L)$ (i.e. maximal non-positive dimension). Such theorems have been proved in the literature (e.g. see **[28]** **[18]** and the references therein). We believe that it will play a fundamental role in further studies of Hamiltonian systems and deserves more attention than it currently does. For the Hamiltonian PDE (2.1) considered in this paper, one important observation is that the operator JL is anti-self-adjoint with respect to the inner product $\langle L\cdot,\cdot\rangle$. Since both anti-self-adjoint and self-adjoint operators are related to unitary operators by the Cayley transform, the Pontryagin invariant subspace theorems can be equivalently stated for unitary, self-adjoint or anti-self-adjoint cases. By Lemma 3.1, e^{tJL} is unitary in $\langle L\cdot,\cdot\rangle$. But to study the eigenvalues of JL more directly, we still use Cayley transform to relate JL to an unitary operator and then apply the Pontryagin invariant subspace theorem. For the sake of completeness, in the following we outline a proof of Theorem 5.1 by the arguments given in [**18**] for the proof of Pontryagin invariant subspace theorem via unitary operators which is based on compactness and fixed point theorems (see also [**25**] [**48**]).

We also give another more constructive proof of Theorem 5.1, by using the Hamiltonian structure of (2.1) and Galerkin approximation. It provides more information about the invariant subspace W.

PROOF. The assumption (2.4) is equivalent to $\ker L = \{0\}$. The first statement of the Theorem follows by the same proof of Lemma A.1. Below we give two different proofs of the construction of the invariant subspace W in the second part of the Theorem.

Proof (**#1.**) Here we sketch a proof of Theorem 5.1 by using the arguments in [**18**]. Let $X_\pm \subset X$ be given by Lemma A.4. Assumptions (**H2-3**) and (2.4) ensure that

$$X = X_- \oplus X_+, \quad X^* = \tilde{X}_-^* \oplus \tilde{X}_+^*, \quad \pm\langle Lu, u\rangle \geq \delta\|u\|^2, \quad \forall u \in X_\pm,$$

where $\tilde{X}_\pm^* = P_\pm^* X_\pm^*$ and $P_\pm$ are the associated projections. As in the proof of Lemma A.5, let $i_{X_\pm} : X_\pm \to X$ be the embedding and

$$L_\pm = \pm P_\pm^* i_{X_\pm}^* L i_{X_\pm} P_\pm, \quad (u, v)_L \triangleq \langle (L_+ + L_-)u, v\rangle.$$

There exists $\delta > 0$ such that $\langle L_\pm u, u\rangle \geq \delta\|u\|^2$, for all $u \in X_\pm$ and the quadratic form $(\cdot,\cdot)_L$ induces an equivalent norm $|u|_L \triangleq \sqrt{(u,v)_L}$ on X. We denote the Hilbert space $\big(X, \langle (L_+ + L_-)\cdot, \cdot\rangle\big)$ by X_L.

Step 1. It is clear that $J(L_+ + L_-)$ is an anti-self-adjoint operator on X_L and JL_- is a bounded linear operator of finite rank on X_L as $L_- X_- = P_-^* X_-^* \subset D(J)$. Writing $JL = J(L_+ + L_-) - 2JL_-$, we obtain that there exists $a > 0$ such that $\alpha \notin \sigma(JL)$ if $|\operatorname{Re}\alpha| \geq a$. Let

$$T = (JL + a)(JL - a)^{-1}, \text{ then } \langle LTu, Tv\rangle = \langle Lu, v\rangle, \ \forall u, v \in X$$

through straightforward calculation using $J = -J^*$ and that L is bounded and symmetric. In some sense, JL is anti-self-adjoint with respect to the quadratic form $\langle L\cdot, \cdot\rangle$ and thus T is formally the Cayley transformation.

Step 2. Let $X_{L\pm}$ be the subspaces $X_\pm$ equipped with the inner product $(\cdot,\cdot)_L$ which is equivalent to $\pm\langle L_{X_\pm}\cdot, \cdot\rangle$ on $X_\pm$, where $L_{X_\pm}$ is defined in (A.1). One may prove (see Lemma 3.6 in [**18**]) that a subspace $W \subset X_L$ satisfies $\dim W = n^-(L)$ and $\langle Lu, u\rangle \leq 0$ for all $u \in W$ if and only if W is the graph of a bounded linear operator $S : X_{L_-} \to X_{L_+}$ with operator norm $|S| \leq 1$. Denote this set of operators, i.e. the unit ball of $L(X_{L-}, X_{L+})$, by $B_1(X_{L-}, X_{L+})$. This proves the first statement.

Step 3. For any $S \in B_1(X_{L-}, X_{L+})$, since T preserves the quadratic form $\langle L\cdot, \cdot\rangle$, one may show that $T\big(\text{graph}(S)\big)$ is still the graph of some $S' \in B_1(X_{L-}, X_{L+})$. Hence we define a transformation $\mathcal{T}$ on $B_1(X_{L-}, X_{L+})$ as

$$\text{graph}\,\big(\mathcal{T}(S)\big) = T\big(\text{graph}(S)\big).$$

Step 4. The space of bounded operators $L(X_{L-}, X_{L+})$ equipped with the weak topology is a locally convex topological vector space. Since X_{L-} is finite dimensional, the unit ball $B_1(X_{L-}, X_{L+})$ is convex and compact under the weak topology. Using the boundedness and the finite dimensionality of X_{L-}, one may prove (see [**18**] for details) that $\mathcal{T}$ is continuous under the weak topology. According to the Tychonoff fixed point theorem (sometimes referred as the Schauder-Tychonoff fixed point theorem, see [**76**]), $\mathcal{T}$ has a fixed point $S \in B_1(X_{L-}, X_{L+})$. Let $W = \text{graph}(S)$ and thus $T(W) \subset W$. According to the definition of T, we have

$$(JL - a)^{-1} = \frac{1}{2a}(T - I)$$

which implies that W is invariant under $(JL-a)^{-1}$. As W is finite dimensional and $(JL - a)^{-1}$ is bounded and injective, it is clear that $W = (JL - a)^{-1}W \subset D(JL)$ and thus $JL(W) \subset W$. □

Alternative proof (#2) of Theorem 5.1 via Galerkin approximation on separable X. On the one hand, the above proof given in [**18**] is elegant and is based on fixed point theorems involving compactness, which does not yield much detailed information of the invariant subspace W. On the other hand, clearly Theorem 5.1 is a generalization into Hilbert spaces of Lemma 4.6 whose constructive proof provides more explicit information of the invariant subspaces. In fact, assuming X is separable, in the rest of this chapter we give an alternative proof of Theorem 5.1 based on Lemma 4.6.

Denote

$$[\cdot, \cdot] = \langle L\cdot, \cdot\rangle \text{ on } X. \tag{5.1}$$

Let $X_\pm$ be the same subspaces of X chosen as in the above proof #1 (as well as in the proof of Proposition A.1). We will study the eigenvalues of JL by a Galerkin approximation. Choose an orthogonal (with respect to $[\cdot, \cdot]$) basis $\{\xi_k\}_{k=1}^{\infty}$ of X such that $\xi_k \in D(JL)$,

$$X_- = span\left\{\xi_1, \cdots, \xi_{n^-(L)}\right\}, \; X_+ = \overline{span\left\{\xi_k\right\}_{k=n^-(L)+1}^{\infty}},$$

and $[\xi_k, \xi_j] = 0$ if $k \neq j$; $[\xi_j, \xi_j] = -1$ if $1 \leq j \leq n^-(L)$; $[\xi_j, \xi_j] = 1$ if $j \geq n^-(L)+1$. For each $n > n^-(L)$, define $X^{(n)} = span\left\{\xi_1, \cdots, \xi_n\right\}$ and denote π^n be the orthogonal projection with respect to the quadratic form $[\cdot, \cdot]$ from X to $X^{(n)}$.

Let X, JL, and $[\cdot, \cdot]$ (as a Hermitian symmetric form) also denote their complexifications as in Chapter 3. Still $\{\xi_1, \xi_2, \ldots\}$ form a basis of the complexified X. Define the operator $F^{(n)} : X^{(n)} \to X^{(n)}$ by

$$F^{(n)}v = \pi^n JLv.$$

Notice that, for $j, k \leq n$,

$$\left[F^{(n)}\xi_k, \xi_j\right] = [\pi^n JL\xi_k, \xi_j] = \langle LJL\xi_k, \xi_j\rangle = \langle L\xi_j, JL\xi_k\rangle \triangleq \left(J^{(n)}\right)_{jk},$$

where the $n \times n$ matrix $\left(J^{(n)}\right)$ is real and anti-symmetric. Let $v = \sum_{j=1}^n y_j \xi_j \in X^{(n)}$ and denote $\vec{y}^{(n)} = (y_1, \cdots, y_n)^T$ and the $n \times n$ matrix

$$H^{(n)} = ([\xi_k, \xi_j]) = diag\left[\underbrace{-1, \cdots, -1}_{1 \text{ to } n^-(L)}, \underbrace{1, \cdots, 1}_{n^-(L)+1 \text{ to } n}\right].$$

Then $F^{(n)}v = \sum_{k=1}^n a_k \xi_k$, where

$$\vec{a}^{(n)} = (a_1, \cdots, a_n)^T = H^{(n)} J^{(n)} \vec{y}^{(n)}.$$

So the eigenvalue problem $F^{(n)}(v) = \lambda v$ is equivalent to

$$H^{(n)} J^{(n)} \vec{y}^{(n)} = \lambda \vec{y}^{(n)}, \tag{5.2}$$

Let $\vec{z}^{(n)} = H^{(n)} \vec{y}^{(n)}$, then the eigenvalue problem (5.2) becomes

$$J^{(n)} H^{(n)} \vec{z}^{(n)} = \lambda \vec{z}^{(n)}.$$

For any $n \geq n^-(L)$, since $n^-\left(H^{(n)}\right) = n^-(L)$, by Lemma 4.6, there exists a subspace $Z^{(n)} \subset \mathbf{C}^n$ of dimension $n^-(L)$, such that $Z^{(n)}$ is invariant under $J^{(n)} H^{(n)}$ and $\left\langle H^{(n)} z, z \right\rangle \leq 0$ for any $z \in Z^{(n)}$. Define $Y^{(n)} = H^{(n)} Z^{(n)}$ and

$$W^{(n)} = \left\{ \sum_{j=1}^{n} y_j \xi_j \mid (y_1, \cdots, y_n)^T \in Y^{(n)} \right\}.$$

Then $W^{(n)}$ is invariant under the linear mapping $F^{(n)}$, $\dim\left(W^{(n)}\right) = n^-(L)$ and the quadratic functions

$$\langle L\cdot, \cdot \rangle |_{W^{(n)}} = \left\langle H^{(n)} \cdot, \cdot \right\rangle |_{Y^{(n)}} = \left\langle H^{(n)} \cdot, \cdot \right\rangle |_{Z^{(n)}} \leq 0.$$

As in the proof (#1) above, denote $P_\pm : X \to X_\pm$ to be the projection operators with $\ker P_\pm = X_\mp$. Since the definitions of X_+ and $W^{(n)}$ imply $W^{(n)} \cap X_+ = \{0\}$, it holds that $P_-\left(W^{(n)}\right) = X_-$. So we can choose a basis $\{w_1^{(n)}, \cdots, w_{n^-(L)}^{(n)}\}$ of $W^{(n)}$ such that $w_j^{(n)} = \xi_j + w_{j+}^{(n)}$ with $w_{j+}^{(n)} \in X_+$. For each $j \leq n^-(L)$, since

$$0 \geq \langle L w_j^{(n)}, w_j^{(n)} \rangle = \langle L \xi_j^{(n)}, \xi_j^{(n)} \rangle + \langle L w_{j+}^{(n)}, w_{j+}^{(n)} \rangle \geq -1 + \delta_0 \|w_{j+}^{(n)}\|^2,$$

so $\|w_j^{(n)}\| \leq C$ for some constant C independent of j and n. Therefore as $n \to \infty$, subject to a subsequence, we have $w_j^{(n)} \rightharpoonup w_j^\infty \in X$ weakly and $P_-(w_j^\infty) = \xi_j$. The subspace $W^\infty = span\left\{w_j^\infty\right\}_{j=1}^{n^-(L)}$ is of dimension $n^-(L)$ since $P_-(W^\infty) = X_-$.

We now show that: i) W^∞ is invariant under the operator JL and ii) $\langle Lu, u \rangle \leq 0$ for any $u \in W^\infty$. To prove i), first note that since $W^{(n)}$ is invariant under $F^{(n)}$, we have

$$F^{(n)} w_k^{(n)} = \sum_{j=1}^{n^-(L)} a_{kj}^{(n)} \; w_j^{(n)}, \quad a_{ij}^{(n)} \in \mathbf{C}.$$

For any integer $l \in \mathbf{N}$ and a fixed $w \in X^{(l)}$, when $n \geq l$,

$$\begin{aligned} \sum_{j=1}^{n^-(L)} a_{kj}^{(n)} \; [w_j^{(n)}, w] &= [F^{(n)} w_k^{(n)}, w] = \langle LJL w_k^{(n)}, w \rangle \\ &= -\langle L w_k^{(n)}, JLw \rangle = -[w_k^{(n)}, JLw]. \end{aligned} \tag{5.3}$$

We claim that $\left\{a_{ij}^{(n)}\right\}$ is uniformly bounded for $1 \leq k, j \leq n^-(L)$ and $n > n^-(L)$. Suppose otherwise, there exists $1 \leq k_0, j_0 \leq n^-(L)$ and a subsequence $\{n_m\} \to \infty$, such that, for all $j \leq n^-(L)$,

$$\left|a_{k_0 j_0}^{(n_m)}\right| = \max_{1 \leq j \leq n^-(L)} \left\{\left|a_{k_0 j}^{(n_m)}\right|\right\} \to \infty \text{ and } \forall j, \; c_{k_0, j} = \lim_{m \to \infty} a_{k_0 j}^{(n_m)} / a_{k_0 j_0}^{(n_m)} \text{ exists }.$$

Then from (5.3), we get

$$\sum_{j=1}^{n^-(L)} \frac{a_{k_0 j}^{(n_m)}}{a_{k_0 j_0}^{(n_m)}} \left[w_j^{(n_m)}, w\right] = -\frac{1}{a_{k_0 j_0}^{(n_m)}} \left[w_{k_0}^{(n_m)}, JLw\right]$$

and letting $m \to \infty$, we obtain

$$\sum_{j=1}^{n^-(L)} c_{k_0,j} \left[w_j^\infty, w\right] = 0 \tag{5.4}$$

where in particular we also notice $|c_{k_0,j_0}| = 1$. By a density argument, the identity (5.4) holds also for any $w \in X$. Therefore $\sum_{j=1}^{n^-(L)} c_{k_0,j}\ w_j^\infty = 0$ by the non-degeneracy of $[\cdot,\cdot]$. This is in contradiction to the independency of $\{w_k^\infty\}$. So $\{a_{kj}^{(n)}\}$ is uniformly bounded. Let $n \to \infty$ in (5.3), subject to a subsequence, we obtain

$$\sum_{j=1}^{n^-(L)} a_{kj}^\infty \left[w_j^\infty, w\right] = -\left[w_j^\infty, JLw\right] = \left[JLw_k^\infty, w\right], \quad \text{where } a_{kj}^\infty = \lim_{n\to\infty} a_{kj}^{(n)}.$$

By a density argument again the above equality is also true for any $w \in X$, which implies

$$JL\left(w_k^\infty\right) = \sum_{j=1}^{n^-(L)} a_{kj}^\infty\ w_j^\infty.$$

So W^∞ is invariant under JL.

Now we prove the above claim ii), that is, $\langle L\cdot,\cdot\rangle|_{W^\infty} \le 0$. For any

$$u = \sum_{j=1}^{n^-(L)} c_j w_j^\infty \in W^\infty.$$

denote

$$u^{(n)} = \sum_{j=1}^{n^-(L)} c_j w_j^{(n)} \in W^{(n)}.$$

Clearly $u^{(n)} \rightharpoonup u$ weakly in X and $\left\langle Lu^{(n)}, u^{(n)}\right\rangle \le 0$, which converges subject to a subsequence. Since

$$\lim_{n\to\infty} \left\langle LP_-u^{(n)}, P_-u^{(n)}\right\rangle = \langle LP_-u, P_-u\rangle,$$

which is due to $P_-w_j^\infty = \xi_j$ and therefore $P_-u^{(n)} \to P_-u$ strongly in X, and

$$\lim_{n\to\infty} \left\langle LP_+u^{(n)}, P_+u^{(n)}\right\rangle \ge \langle LP_+u, P_+u\rangle.$$

as $\langle Lx, x\rangle^{\frac{1}{2}}$ is a norm on X_+. Therefore

$$\begin{aligned} 0 &\ge \lim_{n\to\infty} \left\langle Lu^{(n)}, u^{(n)}\right\rangle = \langle LP_-u, P_-u\rangle + \lim_{n\to\infty} \left\langle LP_+u^{(n)}, P_+u^{(n)}\right\rangle \\ &\ge \langle LP_-u, P_-u\rangle + \langle LP_+u, P_+u\rangle = \langle Lu, u\rangle. \end{aligned} \tag{5.5}$$

This complete the proof of claim ii) and thus the proof of Theorem 5.1 under the separable assumption on X. □

5.2. Further discussions on invariant subspaces and invariant decompositions

Continuous dependence of invariant subspaces on JL. In perturbation problems, the operator JL may depend on a perturbation parameter ϵ. One would naturally wish that a family W_ϵ of non-positive invariant subspaces of dimension $n^-(L)$ may be found depending on ϵ at least continuously. However, this turns out to be impossible in general, even if L is assumed to be non-degenerate. See an example in Section 8.3.

Invariant splitting, I. In the presence of W invariant under JL with $\dim W = n^-(L)$, it is natural to ask whether it is possible to make it into an invariant (under JL) decomposition of X, i.e. whether there exist such W and a codim-$n^-(L)$ invariant subspace $W_1 \subset X$ such that $X = W \oplus W_1$. This is usually not possible as in the following example

$$J = \begin{pmatrix} 0 & -1 & 1 & 0 \\ 1 & 0 & 0 & 1 \\ -1 & 0 & 0 & 0 \\ 0 & -1 & 0 & 0 \end{pmatrix}, \quad L = \begin{pmatrix} 0 & 0 & 0 & 1 \\ 0 & 0 & -1 & 0 \\ 0 & -1 & 0 & 0 \\ 1 & 0 & 0 & 0 \end{pmatrix}, \quad JL = \begin{pmatrix} 0 & -1 & 1 & 0 \\ 1 & 0 & 0 & 1 \\ 0 & 0 & 0 & -1 \\ 0 & 0 & 1 & 0 \end{pmatrix}.$$

Here $n^-(L) = 2$ and the only eigenvalues are $\sigma(JL) = \{\pm i\}$. The only possible non-positive 2-dim invariant subspace, where the eigenvalues of the restriction of JL are contained in $\sigma(JL)$, has to be the geometric kernel of $\pm i$ and thus $W = \{x_3 = x_4 = 0\}$. There does not exist any 2-dim invariant subspace W_1 such that $\mathbf{R}^4 = W \oplus W_1$ since the restriction of JL on W_1 has to have eigenvectors of $\pm i$ as well.

Invariant Splitting, II. In light of Lemma 3.2,

$$W^{\perp_L} = \{u \in X \mid \langle Lu, v\rangle = 0, \ \forall v \in W\}$$

is invariant under e^{tJL}. While one may wish $X = W \oplus W^{\perp_L}$, the only obstacle is that $W^{\perp_L}$ may intersect W nontrivially as L_W, as defined in (A.1), may be degenerate as in the above example. A more natural question is whether it is possible to enlarge W to some closed $\tilde{W} \supset W$ such that

$$\dim \tilde{W} < \infty, \quad JL(\tilde{W}) \subset \tilde{W}, \quad \text{and } L_{\tilde{W}} \text{ an isomorphism.}$$

If so, Lemmas A.2 and 3.2 would imply

$$\tilde{W}^{\perp_L} = \{u \in X \mid \langle Lu, v\rangle = 0, \ \forall v \in \tilde{W}\}$$

is invariant under e^{tJL} and $X = \tilde{W} \oplus \tilde{W}^{\perp_L}$. Moreover $L_{\tilde{W}^{\perp_L}}$ is positive definite due to Theorem 5.1 and thus e^{tJL} is stable on $\tilde{W}^{\perp_L}$. Consequently all the index counting and stability analysis related to e^{tJL} can be reduced to the finite dimensional $\tilde{W}$, which has been analyzed in Chapter 4. For example, we would have a counting theorem like Theorem 2.3, in particular with $k_0^{\leq 0}$ and $k_i^{\leq 0}$ replaced by k_0^- and k_i^-, respectively.

Unfortunately the above splitting is not always possible either, as can be seen from a counterexample in Section 8.4. In Proposition 2.4, we give conditions to get such a decomposition.

CHAPTER 6

Structural decomposition

In this chapter, we prove Theorem 2.1 and Corollary 2.1 on the decomposition of X. Our first step to decompose X is the following proposition based on the invariant subspace Theorem 5.1.

PROPOSITION 6.1. *In addition to* (***H1-H3***), *assume* $\ker L = \{0\}$. *There exist closed subspaces* Y_j, $j = 1, 2, 3, 4$, *such that* $X = \oplus_{j=1}^4 Y_j$ *and*

$$\dim Y_1 = \dim Y_4 = n^-(L) - \dim Y_2 < \infty, \quad Y_{1,2,4} \subset \cap_{k=1}^{\infty} D\big((JL)^k\big), \tag{6.1}$$

and accordingly the linear operator JL *and the quadratic form* $\langle L\cdot, \cdot\rangle$ *take the block forms*

$$JL \longleftrightarrow \begin{pmatrix} \tilde{A}_1 & \tilde{A}_{12} & \tilde{A}_{13} & \tilde{A}_{14} \\ 0 & \tilde{A}_2 & 0 & \tilde{A}_{24} \\ 0 & 0 & \tilde{A}_3 & \tilde{A}_{34} \\ 0 & 0 & 0 & \tilde{A}_4 \end{pmatrix}, \quad L \longleftrightarrow \begin{pmatrix} 0 & 0 & 0 & \tilde{B} \\ 0 & L_{Y_2} & 0 & 0 \\ 0 & 0 & L_{Y_3} & 0 \\ \tilde{B}^* & 0 & 0 & 0 \end{pmatrix}.$$

Here $\tilde{B} : Y_4 \to Y_1^*$ *is an isomorphism and the quadratic forms* $L_{Y_2} \le -\delta_0$ *and* $L_{Y_3} \ge \delta_0$ *for some* $\delta_0 > 0$. *Moreover,* $\tilde{A}_3 : D(\tilde{A}_3) = Y_3 \cap D(JL) \to Y_3$ *is closed, while all other blocks are bounded operators. The operators* $\tilde{A}_{2,3}$ *are anti-self-adjoint with respect to the equivalent inner product* $\mp\langle L_{Y_{2,3}}\cdot, \cdot\rangle$.

Before we give the proof the proposition, we would like to make two remarks. Firstly we observe that $(JL)^k$ takes the same blockwise form as the above one of JL. Secondly, the bounded operator $\tilde{A}_{13}$ should be understood as the closure of $P_1 JL|_{Y_3}$, which may not be closed or everywhere defined itself. Here $P_j : X \to Y_j$ is the projection to Y_j, $j = 1, 2, 3, 4$, according to the decomposition.

PROOF. Theorem 5.1 states that there exists $W \subset D\big((JL)^k\big)$ such that $\dim W = n^-(L)$, $JL(W) \subset W$, and $\langle Lu, u\rangle \le 0$ for all $u \in W$. Let $Y_1 = W \cap W^{\perp_L}$ ($\perp_L$ defined as in Lemma A.2), $\tilde{Y}_2 \subset W$, and $\tilde{Y}_3 \subset W^{\perp_L}$ be closed subspaces such that $W = Y_1 \oplus \tilde{Y}_2$ and $W^{\perp_L} = Y_1 \oplus \tilde{Y}_3$. Recall the notation L_Y as defined in (A.1) for any closed subspace Y.

Claim. $L_{Y_1} = 0$ and there exists $\delta_0 > 0$ such that $L_{\tilde{Y}_2} \le -\delta_0$ and $L_{\tilde{Y}_3} \ge \delta_0$.

In fact, since the quadratic form $\langle Lu, u\rangle \le 0$, for all $u \in W$, the variational principle yields that $u \in W$ satisfies $\langle Lu, u\rangle = 0$ if and only if $\langle Lu, v\rangle = 0$ for all $v \in W$, or equivalently $u \in W \cap W^{\perp_L} = Y_1$. Therefore $\langle Lu, u\rangle < 0$ for any $u \in \tilde{Y}_2 \backslash \{0\}$ which along with $\dim \tilde{Y}_2 < \infty$ implies $L_{\tilde{Y}_2} \le -\delta_0$ for some $\delta_0 > 0$.

If there exists $u \in W^{\perp_L} \backslash W$ satisfying $\langle Lu, u\rangle \le 0$, the definition of $W^{\perp_L}$ would imply $\langle Lu, v\rangle \le 0$ for all $v \in \tilde{W} = W \oplus \mathbf{R}u$ and $\dim \tilde{W} = n^-(L) + 1$. This would contradict Theorem 5.1 and thus we obtain $\langle Lu, u\rangle > 0$ for all $u \in \tilde{Y}_3 \backslash \{0\}$.

Consequently, Lemma A.2 implies that $L_{\tilde{Y}_3} \geq \delta_0$ for some $\delta_0 > 0$ and the claim is proved.

Since L is assumed to non-degenerate, it is easy to see codim-$(W + W^{\perp_L}) = \dim Y_1 < \infty$. Let $\tilde{Y}_4$ be a subspace such that $X = (W + W^{\perp_L}) \oplus \tilde{Y}_4 = Y_1 \oplus \tilde{Y}_2 \oplus \tilde{Y}_3 \oplus \tilde{Y}_4$ and $\tilde{Y}_4 \subset \cap_{k=1}^{\infty} D\big((JL)^k\big)$, which is possible as $\cap_{k=1}^{\infty} D\big((JL)^k\big)$ is dense and $\dim \tilde{Y}_4 = \dim Y_1 < \infty$. With respect to this decomposition, L takes the form

$$L \longleftrightarrow \begin{pmatrix} 0 & 0 & 0 & \tilde{B}_{41}^* \\ 0 & L_{\tilde{Y}_2} & 0 & \tilde{B}_{42}^* \\ 0 & 0 & L_{\tilde{Y}_3} & \tilde{B}_{43}^* \\ \tilde{B}_{41} & \tilde{B}_{42} & \tilde{B}_{43} & L_{\tilde{Y}_4} \end{pmatrix}.$$

The non-degeneracy of L implies that $\tilde{B}_{41} = i_{\tilde{Y}_4}^* L i_{Y_1} : Y_1 \to \tilde{Y}_4^*$ is an isomorphism. Let

$$S_4 = -\frac{1}{2}\tilde{B}_{41}^{-1} L_{\tilde{Y}_4} : \tilde{Y}_4 \to Y_1, \quad S_j = -\tilde{B}_{41}^{-1}\tilde{B}_{4j} : \tilde{Y}_j \to Y_1, \; j = 2, 3.$$

For any $u, v \in \tilde{Y}_4$, we have

$$\begin{aligned} \langle L(u + S_4 u), v + S_4 v\rangle =& \langle L_{\tilde{Y}_4} u, v\rangle + \langle \big(LS_4 + (LS_4)^*\big)u, v\rangle \\ =& \langle L_{\tilde{Y}_4} u, v\rangle + \langle \big(\tilde{B}_{41} S_4 + (\tilde{B}_{41} S_4)^*\big)u, v\rangle = 0. \end{aligned}$$

Similarly, for any $u \in \tilde{Y}_j$, $j = 2, 3$, and $v \in \tilde{Y}_4$,

$$\langle L(u + S_j u), v + S_4 v\rangle = \langle \tilde{B}_{4j} u, v\rangle + \langle \tilde{B}_{41} S_j u, v\rangle = 0.$$

Let $Y_j = (I + S_j)\tilde{Y}_j$. Clearly it still holds $X = \oplus_{j=1}^4 Y_4$. Moreover, $Y_{1,2,4} \subset \cap_{k=1}^{\infty} D\big((JL)^k\big)$, the dimension relationship in (6.1) holds, and in this decomposition L takes the desired form as in the statement of the proposition. Due to $W = Y_1 \oplus Y_2$ and $W^{\perp_L} = Y_1 \oplus Y_3$, the same claim as above implies the uniform positivity of $-L_{Y_2}$ and L_{Y_3}. The non-degeneracy of $\tilde{B}$ follows from the non-degeneracy assumption of L.

The invariance of W, and thus the invariance of $W^{\perp_L}$ due to Lemma 3.2, yields the desired form of JL. The properties that $\tilde{A}_{2,3}$ are anti-self-adjoint with respect to $\langle L_{Y_{2,3}} \cdot, \cdot\rangle$ and the boundedness of other blocks can be proved by applying Lemma A.3 repeatedly to the splitting based on $X = (Y_1 \oplus Y_4) \oplus (Y_2 \oplus Y_3)$. □

The following general functional analysis lemma on invariant subspaces will be used several times in the rest of the paper.

LEMMA 6.1. *Let Z be a Banach space and $Z_{1,2} \subset Z$ be closed subspaces such that $Z = Z_1 \oplus Z_2$. Suppose A is a linear operator on X which, in the above splitting, takes the form* $\begin{pmatrix} A_1 & A_{12} \\ 0 & A_2 \end{pmatrix}$, *such that*

- $A_{1,2} : Z_{1,2} \supset D(A_{1,2}) \to Z_{1,2}$ *are densely defined closed operators, one of which and $A_{12} : Z_2 \to Z_1$ is bounded and*
- $\sigma(A_1) \cap \sigma(A_2) = \emptyset$,

then there exists a bounded operator $S : Z_2 \to Z_1$ such that

(1) $SZ_2 \subset D(A_1)$ *and*
(2) $A\big(\tilde{Z}_2 \cap D(A)\big) \subset \tilde{Z}_2$, *where* $\tilde{Z}_2 = (I + S)Z_2 = \{z_2 + S(z_2) \mid z_2 \in Z_2\}$.

REMARK 6.1. *Clearly the above properties also imply* $D(A)\cap\tilde{Z}_2 = (I+S)D(A_2)$ *is dense in the closed subspace* $\tilde{Z}_2$ *and* $A|_{\tilde{Z}_2} : D(A)\cap\tilde{Z}_2 \to \tilde{Z}_2$ *is a closed operator. Using the splitting* $Z = Z_1 \oplus \tilde{Z}_2$, A *is block diagonalized into* $diag(A_1, A_2)$. *Moreover, if* A_2 *is bounded, then the closed graph theorem implies that* $A|_{\tilde{Z}_2}$ *is also bounded.*

The proof of this lemma may be found in some standard functional analysis textbook. For the sake of completeness we also give a proof here.

PROOF. Let us first consider the case when A_2 is bounded. Since $\sigma(A_2)$ is compact and $\sigma(A_2)\cap\sigma(A_1)=\emptyset$, there exists an open subset $\Omega\subset\mathbf{C}$ with compact closure and smooth boundary $\Gamma=\partial\Omega$ such that $\sigma(A_2)\subset\Omega\subset\overline{\Omega}\subset\mathbf{C}\backslash\sigma(A_1)$. We have

$$\frac{1}{2\pi i}\oint_\Gamma(\lambda-A_1)^{-1}d\lambda=0,\qquad \frac{1}{2\pi i}\oint_\Gamma(\lambda-A_2)^{-1}d\lambda=I.$$

Define

$$S=\frac{1}{2\pi i}\oint_\Gamma T(\lambda)d\lambda,\ \text{where } T(\lambda)=(A_1-\lambda)^{-1}A_{12}(A_2-\lambda)^{-1}.$$

Since $(A_j-\lambda)^{-1}$, $j=1,2$, is analytic from $\mathbf{C}\backslash\sigma(A_j)$ to $L(Z_j)$, it is clear that $S: Z_2\to Z_1$ is bounded. In particular, observing $T(\lambda)z\in D(A_1)$ for any $z\in Z_2$, one may verify

$$T(\lambda)A_2z-A_1T(\lambda)z=(A_1-\lambda)^{-1}A_{12}z-A_{12}(A_2-\lambda)^{-1}z\triangleq\tilde{T}(\lambda)z, \tag{6.2}$$

where $\tilde{T}(\lambda)\in L(Z_2,Z_1)$ is also analytic in λ.

We first show that $Sz\in D(A_1)$ for any $z\in Z_2$. In fact, let S_n, $n\in\mathbf{N}$, be the values of a sequence of Riemann sums of the integral defining S, such that $S_n\to S$. Clearly the discrete Riemann sums satisfy $S_nz\in D(A_1)$ and along with (6.2) we obtain that

$$S_nA_2z-A_1S_nz=\tilde{T}_nz\to\frac{1}{2\pi i}\oint_\Gamma\tilde{T}(\lambda)zd\lambda=A_{12}z$$

where $\tilde{T}_nz$ is the corresponding Riemann sum of the integral on the right side. Therefore we obtain from the closedness of A_1 that $Sz\in D(A_1)$ and

$$SA_2-A_1S=A_{12}. \tag{6.3}$$

From this equation it is straightforward to verify, for any $z\in Z_2$,

$$A(z+Sz)=A_2z+SA_2z. \tag{6.4}$$

In the other case where A_1 is bounded, the proof is similar. In fact, let $\Omega\subset\mathbf{C}$ be an open subset with compact closure and smooth boundary $\Gamma=\partial\Omega$ such that $\sigma(A_1)\subset\Omega\subset\overline{\Omega}\subset\mathbf{C}\backslash\sigma(A_2)$. Define

$$S=-\frac{1}{2\pi i}\oint_\Gamma T(\lambda)d\lambda\in L(Z_2,Z_1).$$

It holds trivially $Sz\in D(A_1)=Z_1$ for any $z\in Z_2$. The same calculation, based on (6.2) but without the need of going through the Riemann sum as $D(A_1)=Z_1$, leads us to (6.3) which implies (6.4) for any $z\in D(A_2)$. The proof is complete. □

In the next step, we remove the non-degeneracy assumption on L and split the phase space X into the direct sum of the hyperbolic (if any) and central subspaces of JL. In particular, the non-degeneracy of the quadratic form $\langle L\cdot,\cdot\rangle$ on the hyperbolic subspace $X_u\oplus X_s$ is of particular importance in the decomposition of JL.

PROPOSITION 6.2. *Assume* **(H1-3)**. *There exist closed subspaces* $X_{u,s,c} \subset X$ *such that*

(1) $X = X_c \oplus X_u \oplus X_s$, $X_{u,s} \subset D(JL)$, $\dim X_u = \dim X_s \leq n^-(L)$, *and* $\ker L \subset X_c$;
(2) *with respect to this decomposition,* JL *and* L *take the forms*

$$JL \longleftrightarrow \begin{pmatrix} A_c & 0 & 0 \\ 0 & A_u & 0 \\ 0 & 0 & A_s \end{pmatrix}, \quad L \longleftrightarrow \begin{pmatrix} L_{X_c} & 0 & 0 \\ 0 & 0 & B \\ 0 & B^* & 0 \end{pmatrix};$$

(3) $B : X_u \to X_s^*$ *is an isomorphism,* A_c *is densely defined, closed, and the spectral sets satisfy* $\sigma(A_c) \subset i\mathbf{R}$ *and* $\pm \operatorname{Re} \lambda > 0$ *for any* $\lambda \in \sigma(A_{u,s})$.

PROOF. Let $X_0 = \ker L$ and $Y = X_- \oplus X_+$ where $X_\pm$ are given in Lemma A.4. Let $P : X \to Y$ be the projection associated to $X = Y \oplus X_0$ and $J_Y = PJP^*$. Lemma A.3 implies that (Y, L_Y, J_Y) satisfy assumptions (**H1-3**), with L_Y being an isomorphism. Applying Proposition 6.1, we obtain closed subspaces Y_j, $j = 1, 2, 3, 4$, such that $X = X_0 \oplus (\oplus_{j=1}^4 Y_j)$ and JL and L take the forms

$$JL \longleftrightarrow \begin{pmatrix} 0 & \tilde{A}_{01} & \tilde{A}_{02} & \tilde{A}_{03} & \tilde{A}_{04} \\ 0 & \tilde{A}_1 & \tilde{A}_{12} & \tilde{A}_{13} & \tilde{A}_{14} \\ 0 & 0 & \tilde{A}_2 & 0 & \tilde{A}_{24} \\ 0 & 0 & 0 & \tilde{A}_3 & \tilde{A}_{34} \\ 0 & 0 & 0 & 0 & \tilde{A}_4 \end{pmatrix}, \quad L \longleftrightarrow \begin{pmatrix} 0 & 0 & 0 & 0 & 0 \\ 0 & 0 & 0 & 0 & \tilde{B} \\ 0 & 0 & L_{Y_2} & 0 & 0 \\ 0 & 0 & 0 & L_{Y_3} & 0 \\ 0 & \tilde{B}^* & 0 & 0 & 0 \end{pmatrix},$$

where $\tilde{B}$ is an isomorphism, $L_{Y_2} \leq -\delta_0$, $L_{Y_3} \geq \delta_0$, for some $\delta_0 > 0$, and $\tilde{A}_{2,3}$ are anti-self-adjoint with respect to the equivalent inner products $\mp\langle L_{Y_{2,3}} \cdot, \cdot \rangle$ on $Y_{2,3}$. The upper triangular structure of JL implies $\sigma(JL) = \{0\} \bigcup \cup_{j=1}^4 \sigma(\tilde{A}_j)$. Moreover we have $\sigma(\tilde{A}_{2,3}) \subset i\mathbf{R}$ due to the anti-self-adjointness of $\tilde{A}_{2,3}$.

For $j = 1, 4$, as $\dim Y_j < \infty$, let $Y_j = Y_{jc} \oplus Y_{jh}$, where Y_{jc} and Y_{jh}, are the eigenspaces of $\tilde{A}_j$ corresponding to all eigenvalues with zero and nonzero real parts, respectively. For any $x_{1,4} \in Y_{1,4}$, the above form of JL and L imply

$$\begin{aligned} \langle \tilde{B}x_4, \tilde{A}_1 x_1 \rangle + \langle \tilde{B}\tilde{A}_4 x_4, x_1 \rangle =& \langle L\tilde{A}_1 x_1, x_4 \rangle + \langle L\tilde{A}_4 x_4, x_1 \rangle \\ =& \langle LJLx_1, x_4 \rangle + \langle LJLx_4, x_1 \rangle = 0. \end{aligned}$$

Much as in the proof of Lemma 3.3, due to the difference in eigenvalues, we obtain

$$\langle \tilde{B}x_{4h}, x_{1c} \rangle = 0 = \langle \tilde{B}x_{4c}, x_{1h} \rangle, \ \forall x_{jc} \in Y_{jc}, \ x_{jh} \in Y_{jh}, \ j = 1, 4. \tag{6.5}$$

Therefore, the non-degeneracy of $\tilde{B}$ implies that

$$\langle \tilde{B}x_{4h}, x_{1h} \rangle \text{ and } \langle \tilde{B}x_{4c}, x_{1c} \rangle \text{ are non-degenerate quadratic forms} \tag{6.6}$$

on $Y_{1h} \times Y_{4h}$ and $Y_{1c} \times Y_{4c}$.

Applying Lemma 6.1 to $X_0 \oplus Y_{1h}$ and $JL|_{X_0 \oplus Y_{1h}}$, we obtain a linear operator $S_1 : Y_{1h} \to X_0$ such that $X_{1h} \triangleq (I + S_1) Y_{1h} \subset D(JL)$ satisfies $JL(X_{1h}) = X_{1h}$ and $\sigma(JL|_{X_{1h}}) = \sigma(\tilde{A}_1|_{Y_{1h}})$. Clearly we still have the decomposition

$$X = X_0 \oplus Y_{1c} \oplus X_{1h} \oplus Y_2 \oplus Y_3 \oplus Y_{4c} \oplus Y_{4h}.$$

Applying again Lemma 6.1 to

$$Z = X_0 \oplus Y_{1c} \oplus Y_2 \oplus Y_3 \oplus Y_{4h}$$

and the projection (with the kernel $X_{1h} \oplus Y_{4c}$) of $JL|_Z$ to Z, we obtain a bounded linear operator

$$S_4 : Y_{4h} \to X_0 \oplus Y_{1c} \oplus Y_2 \oplus Y_3$$

such that $X_{4h} \triangleq (I + S_4)Y_{4h} \subset D(JL)$ satisfies $JL(X_{4h}) \subset X_{1h} \oplus X_{4h}$. Let $X_h = X_{1h} \oplus X_{4h}$, we have

$$X_h \subset D(JL),\ JL(X_h) = X_h,\ \sigma(JL|_{X_h}) = \sigma(JL)\backslash i\mathbf{R} = \sigma(\tilde{A}_1|_{Y_{1h}}) \cup \sigma(\tilde{A}_4|_{Y_{4h}}).$$

According to (6.5) and the form of L, it holds

$$\langle L(I + S_1)x_{1h}, (I + S_4)x_{4h}\rangle = \langle Lx_{1h}, x_{4h}\rangle = \langle \tilde{B}x_{4h}, x_{1h}\rangle.$$

Therefore we obtain the non-degeneracy of $\langle L\cdot,\cdot\rangle$ on X_h from (6.6) and the construction of X_h. Let $X_h = X_u \oplus X_s$, where $X_{u,s}$ are the eigenspaces of all eigenvalues $\lambda \in \sigma(JL|_{X_h})$ with $\pm\mathrm{Re}\lambda > 0$. Lemma 3.3 implies $\langle Lu, v\rangle = 0$ on both X_u and X_s and thus $\langle L\cdot,\cdot\rangle$ is a non-degenerate quadratic form on $X_u \times X_s$ due to the non-degeneracy of $\langle L\cdot,\cdot\rangle$ on X_h. This also yields

$$\dim X_u = \dim X_s = \frac{1}{2}\dim X_h \le \dim Y_1 \le n^-(L).$$

Let

$$X_c = X_h^{\perp_L} = \{u \in X \mid \langle Lu, v\rangle = 0,\ \forall v \in X_h\}.$$

By Lemmas A.2 and 3.2, $X = X_h \oplus X_c$ and X_c is invariant under e^{tJL}. Therefore JL is densely defined on X_c and $A_c\big(D(A_c) \cap X_c\big) \subset X_c$ where $A_c = JL|_{X_c}$. Moreover, from the non-degeneracy of $\langle L\cdot,\cdot\rangle$ on X_h, it is straightforward to show that X_c can be written as a graph of a bounded linear operator from $X_0 \oplus Y_{1c} \oplus Y_2 \oplus Y_3 \oplus Y_{4c}$ to X_h. Therefore, due to the upper triangular structure of JL, the spectrum $\sigma(JL|_{X_c})$ is given by the union of the spectrum of those diagonal blocks of JL complementary to Y_{1h} and Y_{4h} and thus $\sigma(JL|_{X_c}) \subset i\mathbf{R}$. □

As a by-product, we prove the symmetry of eigenvalues of $\sigma(JL)$.

COROLLARY 6.1. *Suppose $\lambda \in \sigma(JL)$.*

(i) If $\lambda \in \sigma(JL)\backslash i\mathbf{R}$, then λ is an isolated eigenvalue of finite algebraic multiplicity. Its eigenspace consists of generalized eigenvectors only. Moreover, let m_λ to be the algebraic multiplicity of λ, then

$$n^-\left(L|_{E_\lambda \oplus E_{-\bar{\lambda}}}\right) = \dim(E_\lambda) = m_\lambda. \tag{6.7}$$

(ii) If λ is an eigenvalues of JL, then $\pm\lambda, \pm\bar{\lambda}$ are also eigenvalues of JL. Moreover, for any integer $k > 0$, $\dim\ker(JL - a)^k$ are the same for $a = \pm\lambda, \pm\bar{\lambda}$.

For an eigenvalue $\lambda \in i\mathbf{R}$, it may happen $\dim\ker(JL - \lambda) = \infty$.

PROOF. According to Lemma 3.6, we only need to prove $\lambda \in \sigma(JL)\backslash i\mathbf{R}$ implies that λ is an isolated eigenvalue of finite multiplicity and $\dim\ker(JL - \lambda)^k = \dim\ker(JL + \bar{\lambda})^k$.

In fact, if $\lambda \in \sigma(JL)\backslash i\mathbf{R}$, then Proposition 6.2 implies that $\lambda \in \sigma(A_u) \cup \sigma(A_s)$. As $A_{u,s}$ are finite dimensional matrices, λ must be an isolated eigenvalue of JL with finite algebraic multiplicity. Moreover, from the blockwise forms of L and JL and $J^* = -J$, it is easy to compute

$$J \longleftrightarrow \begin{pmatrix} J_{X_c} & 0 & 0 \\ 0 & 0 & A_u(B^*)^{-1} \\ 0 & A_sB^{-1} & 0 \end{pmatrix}.$$

Again since $J^* = -J$, we have $A_s = -B^{-1}A_u^*B$. As $A_{u,s}$ are finite dimensional matrices and eigenvalues of JL with positive (or negative) real parts coincide with eigenvalues of A_u (or A_s), the statement in the corollary follows from this similarity immediately.

Since by Proposition 6.2 $L|_{X_u\oplus X_s}$ is non-degenerate, formula (6.7) follows from Lemma 4.2 in the finite dimensional case. □

Proof of Theorem 2.1. Let $X_{5,6} = X_{u,s}$ and $J_{X_c} = P_cJP_c^*$, where $X_{u,s,c}$ are obtained in Proposition 6.2 and $P_c : X \to X_c$ be the projection associated to $X = X_c \oplus X_u \oplus X_s$. According to Lemma A.3, (X_c, L_{X_c}, J_{X_c}) satisfy assumption **(H1-3)** as well. Since Proposition 6.2 also ensures the non-degeneracy of $L_{X_5\oplus X_6}$ and $\dim X_{5,6} \le n^-(L)$, the finite dimensional results in Chapter 4 (Lemma 3.6 and 4.2) imply the symmetry between the spectra $\sigma(A_5)$ and $\sigma(A_6)$ and $n^-(L|_{X_5\oplus X_6}) = \dim X_5$. Therefore we obtain, from the L-orthogonality between X_c and $X_u \oplus X_s$,

$$n^-(L_{X_c}) = n^-(L) - \dim X_5.$$

Recall $X_0 = \ker L = \ker L_{X_c} \subset X_c$. Let $X_\pm$ be given by Lemma A.4 applied to (X_c, L_{X_c}, J_{X_c}), $Y = X_+ \oplus X_-$, $P_Y : X \to Y$ be the associated projection, and $J_Y = P_YJP_Y^*$. Again Lemma A.3 implies (Y, L_Y, J_Y) satisfy **(H1–3)** with L_Y being an isomorphism. Applying Proposition 6.1 to Y and we obtain subspaces $\tilde{X}_j$, $j = 1, 2, 3, 4$. To ensure the orthogonality between $X_0 = \ker L$ and X_j, $j = 1, 2, 3, 4$, we modify the definition of X_j as

$$X_j = \{u \in X_0 \oplus \tilde{X}_j \mid (u, v) = 0, \ \forall v \in \ker L\}, \quad j = 1, 2, 3, 4.$$

It is straightforward to verify the desired properties of the decomposition $X = \oplus_{j=0}^6 X_j$ by using Propositions 6.1 and 6.2. The proof of Theorem 2.1 is complete. □

To finish this chapter, we give the following lemma on the L-orthogonality between certain eigenspaces defined by spectral integrals.

LEMMA 6.2. *Let $\Omega \subset \mathbf{C}$ be an open subset symmetric about $i\mathbf{R}$ with smooth boundary $\Gamma = \partial\Omega$ and compact closure such that $\Gamma \cap \sigma(JL) = \emptyset$. Let*

$$P = \frac{1}{2\pi i}\oint_\Gamma (z - JL)^{-1}dz.$$

and then it holds that $\langle L(I - P)u, Pv\rangle = 0$, for any $u, v \in X$.

The above P is simply the standard spectral projection operator.

PROOF. We first observe for any $w, w' \in X$, (A.8) and (A.10) imply

$$\frac{1}{2\pi i}\oint_\Gamma \langle Lw, (z - JL)^{-1}w'\rangle dz = \langle Lw, Pw'\rangle, \tag{6.8}$$

$$\frac{1}{2\pi i}\oint_\Gamma \langle L(z - JL)^{-1}w, w'\rangle d\bar{z} = -\langle LPw, w'\rangle, \tag{6.9}$$

where the first equality is used in the derivation of the second equality. Here the $d\bar{z}$ and the minus sign in the second equality are due to the anti-linear nature of L in (A.10).

Let $\Omega_1 \subset \Omega$ be an open subset symmetric about $i\mathbf{R}$ such that $\Gamma_1 = \partial\Omega_1 \subset \Omega$ is smooth and $\sigma(JL) \cap (\Omega \backslash \Omega_1) = \emptyset$. Clearly

$$P = \frac{1}{2\pi i} \oint_{\Gamma_1} (z - JL)^{-1} dz,$$

due to the analyticity of $(z - JL)^{-1}$. Denote

$$\tilde{u}(z) = (z - JL)^{-1}u, \; \tilde{v}(z) = (z - JL)^{-1}v, \quad \forall z \notin \sigma(JL).$$

For $z_1, z_2 \notin \sigma(JL)$ satisfying $\bar{z}_1 + z_2 \neq 0$, one may compute using (A.8) and (A.10)

$$\begin{aligned}
&\frac{1}{\bar{z}_1 + z_2}\big(\langle L(z_1 - JL)^{-1}u, v\rangle + \langle Lu, (z_2 - JL)^{-1}v\rangle\big) \\
=&\frac{1}{\bar{z}_1 + z_2}\big(\langle L\tilde{u}(z_1), (z_2 - JL)\tilde{v}(z_2)\rangle + \langle L(z_1 - JL)\tilde{u}(z_1), \tilde{v}(z_2)\rangle\big) \\
=&\langle L\tilde{u}(z_1), \tilde{v}(z_2)\rangle = \langle L(z_1 - JL)^{-1}u, (z_2 - JL)^{-1}v\rangle.
\end{aligned}$$

Due to the definition of Γ_1 and its symmetry about the imaginary axis, $\bar{z}_1 + z_2 \neq 0$ for any $z_1 \in \Gamma$ and $z_2 \in \Gamma_1$. Integrating the above equality along these curves, where Γ_1 is enclosed in Γ, we obtain from the Cauchy integral theorem and (6.8) and (6.9)

$$\begin{aligned}
\langle LPu, Pv\rangle =& \frac{-1}{(2\pi i)^2} \oint_{\Gamma}\oint_{\Gamma_1} \langle L(z_1 - JL)^{-1}u, (z_2 - JL)^{-1}v\rangle dz_2 d\bar{z}_1 \\
=& \frac{-1}{(2\pi i)^2} \oint_{\Gamma}\oint_{\Gamma_1} \frac{1}{\bar{z}_1 + z_2}\langle L(z_1 - JL)^{-1}u, v\rangle dz_2 d\bar{z}_1 \\
&+ \frac{-1}{(2\pi i)^2} \oint_{\Gamma_1}\oint_{\Gamma} \frac{1}{\bar{z}_1 + z_2}\langle Lu, (z_2 - JL)^{-1}v\rangle d\bar{z}_1 dz_2.
\end{aligned}$$

Since $-\bar{z}_1$ is not enclosed in Γ_1 while $-\bar{z}_1$ is enclosed in Γ, the above first integral vanishes and the we obtain from (6.8) and the Cauchy integral theorem

$$\langle LPu, Pv\rangle = \langle Lu, Pv\rangle.$$

This proves the lemma. □

The above lemma implies that $\langle Lu, v\rangle = 0$ for any $u \in \ker P$ and $v \in PX$, where $X = PX \oplus \ker P$ is a spectral decomposition of X invariant under JL. As a corollary, we give the following extension of Lemma 3.3.

Let $\tilde{\sigma} \subset \sigma(JL)$ be compact and also open in the relative topology of $\sigma(JL)$, namely $\tilde{\sigma}$ is isolated in $\sigma(JL)$. There exists an open domain $\Omega \subset \mathbf{C}$ with compact closure and smooth boundary such that $\Omega \cap \sigma(JL) = \tilde{\sigma}$. Let

$$P_{\tilde{\sigma}} = \frac{1}{2\pi i} \oint_{\partial\Omega} (z - JL)^{-1} dz, \quad X_{\tilde{\sigma}} = P_{\tilde{\sigma}}X, \quad X_{\tilde{\sigma}^c} = \ker P_{\tilde{\sigma}}.$$

According to the Cauchy integral theorem, the projection operator $P_{\tilde{\sigma}}$ as well as the above subspaces, which are invariant under JL, are independent of the choice of Ω and $JLP_{\tilde{\sigma}} = P_{\tilde{\sigma}}JL$. Moreover

$$\sigma(JL|_{X_{\tilde{\sigma}}}) = \tilde{\sigma}, \quad \sigma(JL|_{X_{\tilde{\sigma}^c}}) = \sigma(JL)\backslash\tilde{\sigma}.$$

COROLLARY 6.2. *Suppose $\sigma_j \subset \sigma(JL)$, $j = 1, 2$, are compact and also open in the relative topology of $\sigma(JL)$. In addition, assume*

$$\sigma_1 \cap \tilde{\sigma}_2 = \emptyset, \quad \textit{where } \tilde{\sigma}_2 = \{\lambda \in \mathbf{C} \mid \lambda \in \sigma_2 \textit{ or } \bar{\lambda} \in \sigma_2\}.$$

Then $\langle Lu, v\rangle = 0$ *for any* $u \in X_{\sigma_1}$ *and* $v \in X_{\sigma_2}$ *where* $X_{\sigma_{1,2}}$ *are defined as in the above.*

PROOF. According to our assumptions, there exists an open domain $\Omega \subset \mathbf{C}$, symmetric about $i\mathbf{R}$ with smooth boundary and compact closure such that $\Omega \cap \sigma(JL) = \tilde{\sigma}_2$ and $\partial\Omega \cap \sigma(JL) = \emptyset$. The corollary follows from Lemma 6.2 and the facts $X_{\sigma_1} \subset \ker P_{\tilde{\sigma}_2}$ and $X_{\sigma_2} \subset P_{\tilde{\sigma}_2} X$. $\square$

CHAPTER 7

Exponential trichotomy

We prove Theorem 2.2 on the exponential trichotomy in this chapter. The proof is based on the decomposition Theorem 2.1 and we follow the notations there.

Let

$$E^u = X_5, \quad E^s = X_6, \quad E^c = \oplus_{j=0}^4 X_j,$$

where X_j, $j = 0, \ldots, 6$, are given by Theorem 2.1. Based on Theorem 2.1, it only remains to prove the growth estimates.

Since $A_{2,3}$ are anti-self-adjoint with respect to the equivalent inner product $\mp\langle L_{X_{2,3}}\cdot, \cdot\rangle$, there exists a constant $C > 0$ such that

$$|e^{tA_2}|,\ |e^{tA_3}| \le C,\ \forall t \in \mathbf{R}. \tag{7.1}$$

Since $\dim X_5 = \dim X_6 < \infty$ and $\sigma(A_5) = -\sigma(A_6)$, it is clear

$$\begin{aligned} &|e^{tA_5}| \le C(1 + |t|^{\dim X_5 - 1})e^{\lambda_u t},\ \forall t < 0, \\ &|e^{tA_6}| \le C(1 + |t|^{\dim X_6 - 1})e^{-\lambda_u t},\ \forall t > 0 \end{aligned} \tag{7.2}$$

for some $C > 0$ and $\lambda_u = \min\{\mathrm{Re}\lambda \mid \lambda \in \sigma(A_5)\}$. Finally, as $\dim X_1 = \dim X_4 < \infty$ and $\sigma(A_{1,4}) \subset i\mathbf{R}$, we also have

$$|e^{tA_{1,4}}| \le C(1 + |t|^{\dim X_1 - 1}),\ \forall t \in \mathbf{R}. \tag{7.3}$$

For any $x \in X$, write

$$e^{tJL}x = \sum_{j=0}^{6} x_j(t), \quad x_j(t) \in X_j,$$

where X_j, $j = 0, \ldots, 6$, are given by Theorem 2.1. One can write down the equations explicitly:

$$\begin{cases} \partial_t x_0 = A_{01}x_1 + A_{02}x_2 + A_{03}x_3 + A_{04}x_4 \\ \partial_t x_1 = A_1 x_1 + A_{12}x_2 + A_{13}x_3 + A_{14}x_4 \\ \partial_t x_2 = A_2 x_2 + A_{24}x_4 \\ \partial_t x_3 = A_3 x_3 + A_{34}x_4 \\ x_j(t) = e^{tA_j}x_j(0),\ j = 4, 5, 6. \end{cases} \tag{7.4}$$

For $j = 2, 3$, we obtain from Theorem 2.1 and inequalities (7.1) and (7.3) that

$$\begin{aligned} \|x_j(t)\| =& \|e^{tA_j}x_j(0) + \int_0^t e^{(t-\tau)A_j} A_{j4} e^{\tau A_4} x_4(0) d\tau\| \\ \le& C\big(\|x_j(0)\| + (1 + |t|^{\dim X_1})\|x_4(0)\|\big) \end{aligned} \tag{7.5}$$

for some $C > 0$. Regrading $x_1(t)$, we have from (7.1), (7.3), and (7.5)

$$\begin{aligned}\|x_1(t)\| \leq& \|e^{tA_1}x_1(0) + \int_0^t e^{(t-\tau)A_1}\big(A_{12}x_2(\tau) + A_{13}x_3(\tau) + A_{14}e^{\tau A_4}x_4(0)\big)d\tau\| \\ \leq& C\Big(1 + |t|^{\dim X_1 - 1} + \int_0^{|t|} 1 + |t-\tau|^{\dim X_1 - 1}|\tau|^{\dim X_1}d\tau\Big)\|x(0)\| \\ \leq& C(1 + |t|^{2\dim X_1})\|x(0)\|. \end{aligned} \tag{7.6}$$

Much as on the above we also have

$$\|x_0(t)\| \leq C(1 + |t|^{2\dim X_1 + 1})\|x(0)\|, \tag{7.7}$$

The above inequalities prove the desired exponential trichotomy estimates.

Finally, repeatedly applying JL to equation (2.1) and using the above inequalities yield the trichotomy estimates in the graph norms on $D\big((JL)^k\big)$.

CHAPTER 8

The index theorems and the structure of $E_{i\mu}$

Our goal in this chapter is to complete the proof of the index theorems and related properties.

8.1. Proof of Theorem 2.3: the index counting formula

The symmetry of $\sigma(JL)$, the eigenvalues of JL, and the dimensions of the spaces of generalized eigenvectors have been proved in Lemma 3.6 and Corollary 6.1. The index formula (2.13) will be proved in the next two lemmas. Recall the notations $n^-(L|_Y)$ and $n^{\leq 0}(L|_Y)$ for a subspace $Y \subset X$ and indices k_r, k_c, $k^{\leq 0}(i\mu)$, $k_i^{\leq 0}$, $k_0^{\leq 0}$ *etc.* defined in Section 2.4.

LEMMA 8.1. *Under hypotheses **(H1-3)**, it holds*

$$k_r + 2k_c + 2k_i^{\leq 0} + k_0^{\leq 0} \geq n^-(L).$$

PROOF. Let X_j, $j = 0, \ldots, 6$, be the closed subspaces constructed in Theorem 2.1 and $Z = \oplus_{j=0}^{2} X_j$. From Theorem 2.1, Z is an invariant subspace of JL containing $\ker L$ satisfying $\sigma(JL|_Z) \subset i\mathbf{R}$. For any eigenvalue $i\mu \in \sigma(JL|_Z)$, let $E_{i\mu}(Z) = E_{i\mu} \cap Z$ be the subspace of generalized eigenvectors of $i\mu$ in Z, and denote the corresponding non-positive index of $L|_Z$ by

$$k_i^{\leq 0}(Z) = \Sigma_{i\mu \in \sigma(JL|_Z) \cap i\mathbf{R}^+} k^{\leq 0}(i\mu, Z),$$

where $k^{\leq 0}(i\mu, Z) = n^{\leq 0}(L|_{E_{i\mu}(Z)})$.

On the one hand, for any eigenvalue $i\mu \neq 0$, it clearly holds $E_{i\mu}(Z) \subset E_{i\mu}$ and thus $k^{\leq 0}(i\mu, Z) \leq k^{\leq 0}(i\mu)$. Therefore we have $k_i^{\leq 0}(Z) \leq k_i^{\leq 0}$. For the same reason, we also have $k_0^{\leq 0}(Z) \leq k_0^{\leq 0}$ as $\ker L \subset E_0(Z)$, where $k_0^{\leq 0}(Z)$ has a similar definition as $k_0^{\leq 0}$ (defined in (2.12)) except applied to $E_0(Z)$ instead of E_0. From Theorem 2.1 and the finite dimensionality of X_5, it is clear $k_r + 2k_c = \dim X_5$. Consequently we obtain

$$k_r + 2k_c + 2k_i^{\leq 0} + k_0^{\leq 0} \geq \dim X_5 + 2k_i^{\leq 0}(Z) + k_0^{\leq 0}(Z). \tag{8.1}$$

On the other hand, due to the finite dimensionality of X_j, $j = 1, 2$, and the blockwise upper triangular form of JL, we have $Z = \oplus_{i\mu \in \sigma(JL|_Z) \cap i\mathbf{R}} E_{i\mu}(Z)$. Moreover since L is non-positive on Z according to Theorem 2.1, we have

$$2k_i^{\leq 0}(Z) + k_0^{\leq 0}(Z) = \dim X_1 + \dim X_2 = n^-(L) - \dim X_5.$$

Combining it with (8.1), we obtain the conclusion of the lemma. □

LEMMA 8.2. *Under hypotheses **(H1-3)**, it holds*

$$k_r + 2k_c + 2k_i^{\leq 0} + k_0^{\leq 0} \leq n^-(L).$$

PROOF. Let X_j, $j = 0, \dots, 6$, be the closed subspaces constructed in Theorem 2.1 and $Y = \oplus_{j=1}^6 X_j$. Let P_Y be the projection associated to $X = \ker L \oplus Y$. Lemma A.3 implies that (Y, L_Y, J_Y) satisfies assumptions (**H1-3**), where $n^-(L_Y) = n^-(L)$. The definitions of J_Y and L_Y also imply $J_Y L_Y = P_Y(JL)$.

Let $i\mu \in \sigma(JL) \cap i\mathbf{R}^+$. By the definition of $k^{\le 0}(i\mu)$, there exists a subspace $E_{i\mu}^{\le 0} \subset E_{i\mu}$ such that $\dim E_{i\mu}^{\le 0} = k^{\le 0}(i\mu)$ and $\langle Lu, u \rangle \le 0$, for all $u \in E_{i\mu}^{\le 0}$. Since $\mu \ne 0$ and thus $E_{i\mu} \cap \ker L = \{0\}$, we have $\dim P_Y E_{i\mu}^{\le 0} = \dim E_{i\mu}^{\le 0}$. For $\mu < 0$, let $E_{i\mu}^{\le 0} = \{\bar{u} \mid u \in E_{-i\mu}^{\le 0}\}$. For $\mu = 0$, let $\tilde{E}_0 = E_0 \cap Y$ where clearly $E_0 = \ker L \oplus \tilde{E}_0$. There exists a subspace $E_0^{\le 0} \subset \tilde{E}_0$ such that $\dim E_0^{\le 0} = k_0^{\le 0}$ and $\langle Lu, u \rangle \le 0$, for all $u \in E_0^{\le 0}$. Let

$$W = X_5 \oplus E_0^{\le 0} \oplus (\oplus_{i\mu \in \sigma JL \cap i\mathbf{R}} P_Y E_{i\mu}^{\le 0}) \subset Y.$$

It is clearly (the complexification of) a real subspace of Y satisfying $\bar{u} \in W$ for all $u \in W$. Theorem 2.1 implies

$$\dim W = \dim X_5 + k_0^{\le 0} + 2k_i^{\le 0} = k_r + 2k_c + k_0^{\le 0} + 2k_i^{\le 0}.$$

From Lemma 3.3, we have X_5 and $P_Y E_{i\mu}^{\le 0}$ $(i\mu \in \sigma JL \cap i\mathbf{R})$ are mutually L-orthogonal. Therefore our construction of W yields that $\langle Lu, u \rangle \le 0$ for all $u \in W \subset Y$. Applying Theorem 5.1 to (Y, L_Y, J_Y) implies $\dim W \le n^-(L_Y) = n^-(L)$ and thus the lemma is proved. □

8.2. Structures of subspaces $E_{i\mu}$ of generalized eigenvectors

In this section, we will prove Propositions 2.1 and 2.2. We complete the proof in several steps.

LEMMA 8.3. *Let $i\mu \in \sigma(JL) \cap i\mathbf{R}$ and $E \subset E_{i\mu}$ be a closed subspace such that $JL(E) \subset E$. In addition to (**H1-3**), assume $\langle L\cdot, \cdot \rangle$ is non-degenerate (in the sense of (2.4)) on both X and E. Then there exist closed subspaces $E^1, \tilde{E} \subset E$ such that $E = E^1 \oplus \tilde{E}$ and L, JL take the following forms on E*

$$\langle L\cdot, \cdot \rangle \longleftrightarrow \begin{pmatrix} L_{E^1} & 0 \\ 0 & L_{\tilde{E}} \end{pmatrix}, \quad JL \longleftrightarrow \begin{pmatrix} i\mu & 0 \\ 0 & \tilde{A} \end{pmatrix},$$

and $\ker(JL - i\mu) \cap \tilde{E} \subset (JL - i\mu)\tilde{E}$ with non-degenerate L_{E^1} and $L_{\tilde{E}}$ and

$$\dim \tilde{E} \le 3\big(n^-(L|_E) - n^-(L|_{E^1})\big), \ \dim\big((JL - i\mu)E\big) \le 2\big(n^-(L|_E) - n^-(L|_{E^1})\big).$$

REMARK 8.1. *The property $\ker(JL - i\mu) \cap \tilde{E} \subset (JL - i\mu)\tilde{E}$, or equivalently $\ker(\tilde{A} - i\mu) \subset (\tilde{A} - i\mu)\tilde{E}$, is equivalent to that the Jordan canonical form of $\tilde{A}$ contains only nontrivial Jordan blocks.*

PROOF. From Lemma 3.5, $E_{i\mu} = \ker(JL - i\mu)^K$ for some $K > 0$ and $JL : E_{i\mu} \to E_{i\mu}$ is a bounded operator. Let

$$E^0 = \{u \in E \cap \ker(JL - i\mu) \mid \langle Lu, v \rangle = 0, \ \forall v \in E \cap \ker(JL - i\mu)\},$$
$$E^1 = \{u \in E \cap \ker(JL - i\mu) \mid (u, v) = 0, \ \forall v \in E^0\}.$$

Obviously $\ker(JL - i\mu) \cap E = E^0 \oplus E^1$. Moreover for any $u \in E^1 \backslash \{0\}$, there must exist $v \in E^1$ such that $\langle Lu, v \rangle \ne 0$, otherwise it would lead to $u \in E^0$, a contradiction. Applying statement 2 of Lemma A.2 to $Y = E^1$, we obtain that

$\langle L\cdot,\cdot\rangle$ is non-degenerate on E^1. Since $\langle L\cdot,\cdot\rangle$ is assumed to be non-degenerate on both X and E, we apply statement 1 of Lemma A.2 to obtain

$$X = E^1 \oplus (E^1)^{\perp_L} \text{ and } E = E^1 \oplus \tilde{E}, \text{ where } \tilde{E} = E \cap (E^1)^{\perp_L}. \tag{8.2}$$

Here $(E^1)^{\perp_L} \subset X$ is the subspace L-perpendicular to E^1. Clearly $E^0 \subset \tilde{E}$.

Claim. 1.) $\dim \tilde{E} < \infty$, *2.)* $\langle L\cdot,\cdot\rangle$ *is non-degenerate on* $\tilde{E}$, *and 3.)* $JL(\tilde{E}) \subset \tilde{E}$.

The invariance of $\tilde{E}$ under JL follows directly from the invariance of E and E^1 and Lemma 3.2. The non-degeneracy of $\langle L\cdot,\cdot\rangle$ on both E and E^1 implies that $\langle L\cdot,\cdot\rangle$ is non-degenerate on $\tilde{E}$ as well. To complete the proof of the claim, we only need to prove $\dim \tilde{E} < \infty$.

On the one hand, from the above definitions, $\langle Lu, v\rangle = 0$ for any $u, v \in E^0$. The non-degeneracy of $\langle L\cdot,\cdot\rangle$ on E and Theorem 5.1 along with Remark 5.1 imply

$$\dim E^0 \le n^-(L|_{\tilde{E}}). \tag{8.3}$$

On the other hand, it is clear from the definitions of $\tilde{E}$ and the non-degeneracy of $\langle L\cdot,\cdot\rangle$ on E^1 that

$$\tilde{E} \cap \ker(JL - i\mu) = E^0. \tag{8.4}$$

Moreover, from Lemma 3.5, $E \subset E_{i\mu} = \ker(JL - i\mu)^K$ for some $K > 0$, and each Jordan chain in $\tilde{E}$ contains a vector in E^0, we obtain

$$\dim \tilde{E} \le K \dim E^0 \le K n^-(L|_E),$$

from the invariance of $\tilde{E}$ under JL. The claim is proved.

Now we complete the proof of the lemma by reducing it to a finite dimensional problem satisfying our framework. Firstly, to replace $\tilde{E}$ by the complexification of some real Hilbert space, let

$$\tilde{E}^R = \{u + \bar{v} \mid u, v \in \tilde{E}\}$$

which satisfies $\bar{u} \in \tilde{E}^R$ for any $u \in \tilde{E}^R$. Since $u \in E_{i\mu}$ implies $\bar{u} \in E_{-i\mu}$, we have $\tilde{E}^R = \tilde{E}$ if $\mu = 0$. If $\mu \neq 0$, from (A.12) and Lemma 3.3 we obtain

$$\langle L\bar{u}, v\rangle = 0, \ \langle L\bar{u}, \bar{v}\rangle = \overline{\langle Lu, v\rangle}, \quad \forall u, v \in \tilde{E}.$$

Therefore $\tilde{E}^R$ satisfies the same properties as in the above claim whether $\mu = 0$ or not. Using the non-degeneracy of $\langle L\cdot,\cdot\rangle$ on X and $\tilde{E}^R$, and applying Lemma A.3 to the splitting $X = \tilde{E}^R \oplus (\tilde{E}^R)^{\perp_L}$ with the associated projections $P_{\tilde{E}^R}$ and $I - P_{\tilde{E}^R}$, we have that the combination $(\tilde{E}^R, L_{\tilde{E}^R}, J_{\tilde{E}^R})$ satisfies assumptions (**H1-3**), where $J_{\tilde{E}^R} = P_{\tilde{E}^R} J P_{\tilde{E}^R}^*$. We may apply Proposition 2.2, whose finite dimensional case under the non-degeneracy assumption on $\langle L\cdot,\cdot\rangle$ has been proved in Chapter 4. As $\tilde{E} = \tilde{E}^R \cap \ker(JL - i\mu)^K$, that canonical form implies

$$\dim\big((JL - i\mu)\tilde{E}\big) \le 2n^-(L|_{\tilde{E}}), \quad \ker(JL - i\mu) \cap \tilde{E} \subset (JL - i\mu)\tilde{E},$$

where (8.4) is also used along with the canonical form. We notice $(JL - i\mu)E = (JL - i\mu)\tilde{E}$, as $E^1 \subset \ker(JL - i\mu)$, and thus

$$\dim\big((JL - i\mu)E\big) \le 2n^-(L|_{\tilde{E}}) = 2\big(n^-(L|_E) - n^-(L_{E^1})\big).$$

The block forms of L and JL follow from the L-orthogonality and the invariance of the splitting $E = E^1 \oplus \tilde{E}$. Finally the estimate on $\dim \tilde{E}$ follows from the above inequality and (8.3) and (8.4). The proof is complete. $\square$

Next we study $E_{i\mu}$ by assuming the non-degeneracy of L.

LEMMA 8.4. *In addition to (**H1-3**), assume $\langle L\cdot,\cdot\rangle$ is non-degenerate. Let $i\mu \in \sigma(JL)\cap i\mathbf{R}$. There exist subspaces $E^{D,1,G} \subset E_{i\mu}$ such that*

$$E_{i\mu} = E^D \oplus E^1 \oplus E^G, \quad \dim\big((JL - i\mu)E_{i\mu}\big) \le 2\big(k^{\le 0}(i\mu) - n^-(L|_{E^1})\big),$$

$$\dim E^G \le 3\big(k^{\le 0}(i\mu) - \dim E^D - n^-(L|_{E^1})\big),$$

and L and JL take the block forms on E

$$\langle L\cdot,\cdot\rangle \longleftrightarrow \begin{pmatrix} 0 & 0 & 0 \\ 0 & L_1 & 0 \\ 0 & 0 & L_G \end{pmatrix}, \quad JL \longleftrightarrow \begin{pmatrix} A_D & A_{D1} & A_{DG} \\ 0 & i\mu & 0 \\ 0 & 0 & A_G \end{pmatrix}$$

where all blocks are bounded operators and L_1 and L_G are non-degenerate. Moreover $\ker(A_G - i\mu) \subset (A_G - i\mu)E_G$.

PROOF. Again, to apply previous results directly it would be easier to consider the complexifications of real Hilbert spaces

$$I_{i\mu} \triangleq E_{i\mu} + E_{-i\mu} = \{u + \bar{v} \mid u, v \in E_{i\mu}\}.$$

Due to Lemma 3.5, $JL|_{I_{i\mu}}$ is bounded with

$$L(I_{i\mu}) \subset D(J), \; JL(I_{i\mu}) \subset I_{i\mu}, \; \sigma(JL|_{I_{i\mu}}) = \{\pm i\mu\}.$$

We split the spaces by starting with

$$E^D = \{u \in E_{i\mu} \mid \langle Lu, v\rangle = 0, \; \forall v \in E_{i\mu}\}, \; I^D = \{u + \bar{v} \mid u, v \in E^D\},$$

$$E^{ND}_{i\mu} = \{u \in E_{i\mu} \mid (u, v) = 0, \; \forall v \in E^D\}, \; I^{ND} = \{u + \bar{v} \mid u, v \in E^{ND}_{i\mu}\}.$$

From the anti-symmetry of JL with respect to $\langle L\cdot,\cdot\rangle$ and the invariance of $E_{i\mu}$ along with (A.12), we have $JL(I^D) \subset I^D$. In the splitting $I_{i\mu} = I^D \oplus I^{ND}$, $\langle L\cdot,\cdot\rangle$ and JL can be represented in the following block forms

$$\langle L\cdot,\cdot\rangle \longleftrightarrow \begin{pmatrix} 0 & 0 \\ 0 & L_{ND} \end{pmatrix}, \quad JL \longleftrightarrow \begin{pmatrix} A_D & A_{D,ND} \\ 0 & A_{ND} \end{pmatrix},$$

where all blocks are bounded real (satisfying (A.12)) operators. In particular $\ker(L|_{I_{i\mu}}) = I^D$ and $I_{i\mu} = I^D \oplus I^{ND}$ and thus Lemma A.2 implies that $L_{ND} : I^{ND} \to (I^{ND})^*$ is an isomorphism. The anti-symmetry of JL with respect to $\langle L\cdot,\cdot\rangle$ yields $L_{ND}A_{ND} + A_{ND}^* L_{ND} = 0$. Therefore

$$J_{ND} = A_{ND}L_{ND}^{-1} : (I^{ND})^* \to I_{ND}$$

is an anti-symmetric bounded operator satisfying $A_{ND} = J_{ND}L_{ND}$. Clearly the combination (I^{ND}, L_{ND}, J_{ND}) satisfies (**H1-3**) with the non-degenerate L_{ND}. Moreover, $\sigma(J_{ND}L_{ND}) = \{\pm i\mu\}$ with the eigenspace of $i\mu$ given by $E^{ND}_{i\mu}$ where $\langle L_{ND}\cdot,\cdot\rangle$ is also non-degenerate. Therefore we may apply Lemma 8.3 (with X and E replaced by I^{ND} and $E^{ND}_{i\mu}$, respectively) to obtain the splitting $E^{ND}_{i\mu} = E^1 \oplus E^G$ and the desired block forms of L and JL follow. The desired estimate on $\dim E^G$ is obtained by noting

$$k^{\le 0}(i\mu) = n^-(L|_{E^{ND}_{i\mu}}) + \dim E^D. \tag{8.5}$$

Moreover, according to Lemma 8.3, we have

$$\dim(A_{ND} - i\mu)E^{ND}_{i\mu} \le 2\big(n^-(L|_{E^{ND}_{i\mu}}) - n^-(L|_{E^1})\big).$$

Along with (8.5) and the block form of JL, it implies

$$\begin{aligned}\dim(JL - i\mu)E_{i\mu} &\le \dim E^D + \dim(A_{ND} - i\mu)E_{i\mu}^{ND}\\ &\le \dim E^D + 2\big(n^-(L|_{E_{i\mu}^{ND}}) - n^-(L|_{E^1})\big) \le 2\big(k^{\le 0}(i\mu) - n^-(L|_{E^1})\big)\end{aligned}$$

which finishes the proof. □

Proof of Proposition 2.1 and Proposition 2.2. What remains to be proved in these two propositions can be obtained in a similar framework and we complete their proofs together here.

Let $X_\pm$ be given by Lemma A.4 and $X_1 = X_- \oplus X_+$. Clearly $X = X_0 \oplus X_1$, where $X_0 = \ker L$, with the associated projections $P_{X_{0,1}}$. According to Lemma A.3, $\langle L\cdot, \cdot\rangle$ and JL take the following block forms

$$\langle L\cdot,\cdot\rangle \longleftrightarrow \begin{pmatrix} 0 & 0 \\ 0 & L_{X_1}\end{pmatrix}, \quad JL \longleftrightarrow \begin{pmatrix} 0 & A_1 \\ 0 & J_{X_1}L_{X_1}\end{pmatrix},$$

where $A_1 : X_1 \to \ker L$ is bounded and $L_{X_1} = i_{X_1}^* L i_{X_1} : X_1 \to X_1^*$ and $J_{X_1} = P_{X_1} J P_{X_1}^*$. Moreover, Lemmas A.3 and A.4 imply that (X_1, L_{X_1}, J_{X_1}) satisfies assumptions (**H1-3**) with the isomorphic L_{X_1} and $n^-(L_{X_1}) = n^-(L)$. For any eigenvalue $i\mu \in i\mathbf{R}$, let $E_{i\mu}^1$ be the subspace of generalized eigenvectors of $i\mu$ for $J_{X_1}L_{X_1}$, possibly $\{0\}$ if $\mu = 0$. From Lemma 3.5 and 8.4, for some $K > 0$,

$$E_{i\mu}^1 = \ker(J_{X_1}L_{X_1} - i\mu)^K, \quad \dim(J_{X_1}L_{X_1} - i\mu)E_{i\mu}^1 \le 2n^{\le 0}(L_{X_1}|_{E_{i\mu}^1}).$$

For any integer $k > 0$, $(JL - i\mu)^k$ takes the block form

$$(JL - i\mu)^k \longleftrightarrow \begin{pmatrix} (-i\mu)^k & A_k \\ 0 & (J_{X_1}L_{X_1} - i\mu)^k\end{pmatrix},$$

where the linear operator $A_k : X_1 \to \ker L$ can be computed inductively

$$A_{k+1} = (-i\mu)^k A_1 + A_k(J_{X_1}L_{X_1} - i\mu), \quad D\big((J_{X_1}L_{X_1} - i\mu)^k\big) \subset D(A_{k+1}).$$

It is straightforward to show

$$(8.6) \qquad u \in E_{i\mu} \iff P_{X_1}u \in E_{i\mu}^1 \text{ and } (-i\mu)^K P_{X_0}u + A_K P_{X_1}u = 0.$$

We first consider $\mu \ne 0$. We obtain from (8.6)

$$E_{i\mu} = \{u - (-i\mu)^{-K}A_K u \mid u \in E_{i\mu}^1\},$$

i.e. vectors in $E_{i\mu}$ are determined only by their X_1-component. From Lemma 3.5 and Remark 3.1, $E_{i\mu}^1$ and $E_{i\mu}$ are both subspaces. Therefore A_K is a bounded operator. Since

$$\langle L\big(u - (-i\mu)^{-K}A_K u\big), v - (-i\mu)^{-K}A_K v\rangle = \langle Lu, v\rangle, \ \forall u, v \in E_{i\mu}^1,$$

we obtain from Lemma 8.4

$$\begin{aligned}\dim(JL - i\mu)E_{i\mu} &= \dim(J_{X_1}L_{X_1} - i\mu)E_{i\mu}^1\\ &\le 2n^{\le 0}(L_{X_1}|_{E_{i\mu}^1}) = 2n^{\le 0}(L_{E_{i\mu}}) = 2k^{\le 0}(i\mu).\end{aligned}$$

This proves the desired estimate on $\dim(JL - i\mu)E_{i\mu}$ in Proposition 2.1. Along with Lemma 3.5, it completes the proof of Proposition 2.1 in the case of $\mu \ne 0$.

To prove Proposition 2.2, let $E_{i\mu}^1 = \tilde{E}^D \oplus \tilde{E}^1 \oplus \tilde{E}^G$ where these subspaces are given by Lemma 8.4 for $J_{X_1}L_{X_1}$. Let

$$E^{D,1,G} = \{u - (-i\mu)^{-K}A_K u \mid u \in \tilde{E}^{D,1,G}\}.$$

It is easy to verify that they satisfy the properties in Proposition 2.2. Since $\dim E^G < \infty$, the 'good' basis of E^G has been constructed in the finite dimensional cases in Chapter 4 and the proof of Proposition 2.2 is complete.

For $\mu = 0$, it is easy to see from the above block forms

$$E_0 = X_0 \oplus E_0^1.$$

Therefore we have

$$(JL)^2 E_0 = (JL)^2 E_0^1 = JL(P_{X_1} JLE_0^1) = JL(J_{X_1} L_{X_1} E_0^1),$$

which along with Lemma 8.4 implies

$$\dim\ (JL)^2 E_0 \le \dim\ J_{X_1} L_{X_1} E_0^1 \le 2k_0^{\le 0}.$$

This completes the proof of Proposition 2.1 in the case of $\mu = 0$.

To prove Proposition 2.2, let $E_0^1 = E^D \oplus E^1 \oplus E^G$ where these subspaces are given by Lemma 8.4 for $J_{X_1} L_{X_1}$ and $\mu = 0$. It is easy to verify that they satisfy the properties in Proposition 2.2. Again since $\dim E^G < \infty$, the 'good' basis of E^G has been constructed in the finite dimensional cases in Chapter 4 and the proof of Proposition 2.2 is complete. $\square$

REMARK 8.2. *In the case of $\mu = 0$, we can not replace $(JL)^2 E_0$ by JLE_0, as seen from the following counterexample. Consider $X = Y \oplus Y \oplus \mathbf{R^2}$ where Y is any Hilbert space. Let*

$$J = \begin{pmatrix} 0 & I & 0 & 0 \\ -I & 0 & 0 & 0 \\ 0 & 0 & 0 & -1 \\ 0 & 0 & 1 & 0 \end{pmatrix}, \quad L = \begin{pmatrix} 0 & 0 & 0 & 0 \\ 0 & I & 0 & 0 \\ 0 & 0 & 1 & 0 \\ 0 & 0 & 0 & -1 \end{pmatrix}, \quad JL = \begin{pmatrix} 0 & I & 0 & 0 \\ 0 & 0 & 0 & 0 \\ 0 & 0 & 0 & 1 \\ 0 & 0 & 1 & 0 \end{pmatrix}.$$

It is clear that $k_0^{\le 0} = 0$, $\ker L = X_0 = Y \oplus \{0\} \oplus \{(0,0)^T\}$, $E_0 = Y \oplus Y \oplus \{(0,0)^T\}$, and $\dim JLE_0 = \dim \ker L = \dim Y$.

8.3. Subspace of generalized eigenvectors E_0 and index $k_0^{\le 0}$

In this Section we prove Propositions 2.7, 2.8, Lemma 2.1 and Corollary 2.3, 2.4 on the subspace E_0 and the non-positive index $k_0^{\le 0}$ for the eigenvalue 0.

Proof of Proposition 2.7. According to Corollary A.1, $LJ : D(J) \to X$ is closed and thus LJL is also closed. Therefore, $(JL)^{-1}(\ker L) = \ker(LJL)$ is also closed.

Since $(JL)^{-1}(\ker L) \subset E_0$, due to the hyperbolicity of JL on $X_{5,6}$, we have $(JL)^{-1}(\ker L) \subset \oplus_{j=0}^4 X_j$, where the decomposition of $X = \ker L \oplus \oplus_{j=1}^6 X_j$ is given in Theorem 2.1. Let

$$S = (JL)^{-1}(\ker L) \cap \oplus_{j=1}^4 X_j.$$

Since $X_0 = \ker L \subset (JL)^{-1}(\ker L)$, we have $\ker L \oplus S = (JL)^{-1}(\ker L) \subset E_0$. Therefore, from the definition of $k_0^{\le 0}$ it is clear $k_0^{\le 0} \ge n_0 = n^{\le 0}(L|_S)$ and we only need to prove (ii) of Proposition 2.7.

Assume in addition that $\langle L\cdot, \cdot\rangle$ is non-degenerate on $(JL)^{-1}(\ker L)/\ker L$. We claim

$$E_0 = (JL)^{-1}(\ker L). \tag{8.7}$$

In fact, suppose $u \in E_0 \backslash \big((JL)^{-1}(\ker L)\big)$. There exists $m > 0$ such that

$$u_1 \triangleq (JL)^{m-1} u \notin (JL)^{-1}(\ker L)$$
$$u_0 \triangleq JLu_1 = (JL)^m u \in (JL)^{-1}(\ker L) \backslash \ker L.$$

It follows that, for any $v \in (JL)^{-1}(\ker L)$,

$$JLv \in \ker L \Longrightarrow \langle Lu_0, v\rangle = \langle L(JL)u_1, v\rangle = -\langle Lu_1, JLv\rangle = 0.$$

The existence of such u_0 would imply $\langle L\cdot,\cdot\rangle$ is degenerate on $(JL)^{-1}(\ker L)/\ker L$, contradictory to our assumption. Therefore (8.7) is proved and consequently we obtain from the definition of $k_0^{\le 0}$ that

$$k_0^{\le 0} = n^{\le 0}\big(\langle L\cdot,\cdot\rangle|_{(JL)^{-1}(\ker L)/\ker L}\big) = n^{-}\big(\langle L\cdot,\cdot\rangle|_{(JL)^{-1}(\ker L)/\ker L}\big)$$

due to the non-degeneracy assumption. This completes the proof of the proposition. □

We will prove Lemma 2.1, Proposition 2.8, and Corollary 2.3 and 2.4 in the rest of the section. We first observe that it is straightforward to show $\langle Lu, v\rangle = 0$, for any $u \in \ker(JL)$ and $v \in R(J)$. Through a density argument, we obtain

$$\langle Lu, v\rangle = 0, \quad \forall u \in \ker(JL),\ v \in \overline{R(J)}. \tag{8.8}$$

Throughout the rest of this section, let $S_1, S_2, S^\#$ be defined as in Corollaries 2.3 and 2.4, i.e.

$$\ker(JL) = \ker L \oplus S_1, \qquad \overline{R(J)} = \big(\overline{R(J)} \cap \ker L\big) \oplus S^\#$$

and

$$\overline{R(J)} \cap (JL)^{-1}(\ker L) = S_2 \oplus \big(\overline{R(J)} \cap \ker L\big).$$

LEMMA 8.5. *Suppose $\langle L\cdot,\cdot\rangle$ is non-degenerate on $S^\#$, then it is also non-degenerate on S_1 and moreover,*

$$X = \ker(JL) \oplus S^\# = \ker L \oplus S_1 \oplus S^\#. \tag{8.9}$$

PROOF. The non-degeneracy of $\langle L\cdot,\cdot\rangle$ on $S^\#$ implies the non-degeneracy of $L_{S^\#} : S^\# \to (S^\#)^*$, which is defined in (A.1). For any $u \in X$, as in the proof of Lemma A.2, let

$$u^\# = L_{S^\#}^{-1} i_{S^\#}^* Lu \in S^\#$$

which satisfies

$$\langle Lu_1, v\rangle = 0,\ \forall v \in S^\#, \quad \text{where } u_1 = u - u^\#.$$

By the definition of $S^\#$, we also have

$$\langle Lu_1, v\rangle = 0,\ \forall v \in \overline{R(J)}.$$

Since $J^* = -J$, we obtain

$$Lu_1 \in \ker J^* = \ker J \Longrightarrow u_1 \in \ker(JL) = \ker L \oplus S_1.$$

Therefore $u = u_1 + u^\# \in \ker(JL) + S^\#$ and thus $X = \ker(JL) + S^\#$.

For any $u \in S^\# \cap ker(JL)$, from (8.8) we obtain $\langle Lu, v\rangle = 0$, for any $v \in \ker(JL) + \overline{R(J)} \supset \ker(JL) + S^\# = X$. Therefore $u \in \ker L$. Since $u \in S^\# \cap \ker L = \{0\}$, we have $u = 0$ and thus $X = \ker(JL) \oplus S^\# = \ker L \oplus S_1 \oplus S^\#$.

From Lemma A.2, $\langle L\cdot,\cdot\rangle$ is non-degenerate on $S^\# \oplus S_1$. Since it is also assumed to be non-degenerate on $S^\#$, the non-degeneracy of $\langle L\cdot,\cdot\rangle$ on S_1 follows from the L-orthogonality (8.8) between S_1 and $S^\#$. □

LEMMA 8.6. *Suppose $\langle L\cdot,\cdot\rangle$ is non-degenerate on S_1, then it is also non-degenerate on $S^\#$.*

PROOF. Like in the proof of the previous lemma, the non-degeneracy of $\langle L\cdot,\cdot\rangle$ on S_1 implies the non-degeneracy of $L_{S_1}: S_1 \to S_1^*$. For any $u \in X$, as in the proof of Lemma A.2, let

$$u_1 = L_{S_1}^{-1} i_{S_1}^* Lu \in S_1$$

which satisfies

$$\langle Lu_*, v\rangle = 0, \ \forall v \in \ker(JL) = \ker L \oplus S_1, \quad \text{where } u_* = u - u_1.$$

Since $JL = -(LJ)^*$, we obtain $Lu_* \in \overline{R(LJ)}$.

Claim: $\overline{R(LJ)} = L(S^\#)$. In fact, it is easy to see $L(S^\#) \subset L\big(\overline{R(J)}\big) \subset \overline{R(LJ)}$ due to the boundedness of L. In the following we will prove that $\overline{R(LJ)} \subset L(S^\#)$. Let $y \in \overline{R(LJ)}$, there exists a sequence $y_n = LJx_n$ such that $y_n \to y$ as $n \to +\infty$. Since $\overline{R(J)} = \ker L \oplus S^\#$, let $Jx_n = z_{n,0} + z_{0,\#}$ where $z_{n,0} \in \ker L$ and $z_{n,\#} \in S^\#$. As $y_n = LJx_n = Lz_{n,\#} \to y$ and the non-degeneracy assumption of $\langle L\cdot,\cdot\rangle$ on $S^\#$ implies that $L|_{S^\#}: S^\# \to L(S^\#)$ is an isomorphism, we obtain that $\{z_{n,\#}\}$ is a Cauchy sequence. Let $z_{n,\#} \to z_\# \in S^\#$ and then $y = Lz_\# \in L(S^\#)$. The claim is proved.

We can now finish the proof of the lemma. Since we have proved

$$L(u - u_1) = Lu_* \in \overline{R(LJ)} = L(S^\#),$$

there exists $u_\# \in S^\#$ such that $L(u-u_1) = Lu_\#$. Let $u_0 = u - u_1 - u_\#$. Clearly $u_0 \in \ker L$. Therefore $u = u_0 + u_1 + u_\#$ and thus $X = \ker L \oplus S_1 \oplus S^\# = \ker(JL) \oplus S^\#$. The proof of $\ker(JL) \cap S^\# = \{0\}$ and consequently the non-degeneracy of $\langle L\cdot,\cdot\rangle$ on S_1 is the same as in the proof of the last lemma. □

The conclusion in Lemma 2.1 is already contained in the above lemmas.

Proof of Proposition 2.8 and equivalently Corollary 2.4. The property $X = \ker(JL) + \overline{R(J)}$ is a direct consequence of (8.9). Along with (8.8), it also implies $\overline{R(J)} \cap \ker(JL) = \overline{R(J)} \cap \ker L \subset \ker L$.

From the L-orthogonality (8.8), the decomposition (8.9), and the non-degeneracy of $\langle L\cdot,\cdot\rangle$ on $S_1, S^\#$, and $S_1 \oplus S^\#$, we immediately obtain $n^- = n^-(L|_{S_1}) + n^-(L|_{S^\#})$.

From the decomposition (8.9) and the definitions of S_1 and S_2, we have

$$(JL)^{-1}\ker L = \ker L \oplus S_1 \oplus S_2.$$

Therefore $k_0^{\leq 0} \geq n^-(L|_{S_1}) + n^{\leq 0}(L|_{S_2})$ follows from Proposition 2.7.

Finally, let us assume, in addition, that $\langle L\cdot,\cdot\rangle$ is non-degenerate on S_2. Immediately we have the non-degeneracy of $\langle L\cdot,\cdot\rangle$ on $(JL)^{-1}(\ker L)/\ker L$ and Proposition 2.7 implies $k_0^{\leq 0} = n^-(L|_{S_1}) + n^{\leq 0}(L|_{S_2})$. The proof is complete. □

8.4. Non-degeneracy of $\langle L\cdot,\cdot\rangle$ on $E_{i\mu}$ and isolated purely imaginary spectral points

In Proposition 2.2, the presence of the subspace $E^D \subset E_{i\mu}$ is due to the possible degeneracy of $\langle L\cdot,\cdot\rangle$ on $E_{i\mu}$. Otherwise the statement of the proposition would be much more clean and some results can be improved. However, in case when $i\mu$ is not isolated in $\sigma(JL)$, it is indeed possible that $\langle L\cdot,\cdot\rangle$ degenerates on $E_{i\mu}$ even if it is non-degenerate on X.

Example of degenerate $\langle L\cdot,\cdot\rangle$ on $E_{i\mu}$. Consider $X = \mathbf{R}^{2n} \oplus \mathbf{R}^{2n} \oplus X_1$, where X_1 is a Hilbert space. Here we identify Hilbert spaces and their dual spaces via Riesz Representation Theorem. Let $\mu \in \mathbf{R}$ and

• $A : X_1 \supset D(A) \to X_1$ be an anti-self-adjoint operator such that $i\mu \in \sigma(A)$ is *not* an eigenvalue;

• $A_1 : \mathbf{R}^{2n} \to X_1$ such that $\ker A_1 = \{0\}$ and, after the complexification of A and A_1 into complex linear operators, $R(A_1) \cap R(A \pm i\mu) = \{0\}$, which is possible due to the spectral assumption on A; and

• $J = \begin{pmatrix} 0 & J_{2n} & 0 \\ J_{2n} & J_{2n} & -B^{-1}A_1^* \\ 0 & A_1 B^{-1} & A \end{pmatrix}$, $L = \begin{pmatrix} 0 & B & 0 \\ B & 0 & 0 \\ 0 & 0 & I_{X_1} \end{pmatrix}$, where $B_{2n\times 2n}$ is any symmetric matrix and $J_{2n} = \begin{pmatrix} 0 & -I_{n\times n} \\ I_{n\times n} & 0 \end{pmatrix}$.

One may compute

$$JL = \begin{pmatrix} J_{2n}B & 0 & 0 \\ J_{2n}B & J_{2n}B & -B^{-1}A_1^* \\ A_1 & 0 & A \end{pmatrix}.$$

LEMMA 8.7. *For any integer $k > 0$,*

$$\ker(JL - i\mu)^k = \{(0, x, 0)^T \mid x \in \ker(J_{2n}B - i\mu)^k\} \subset \mathbf{R}^{2n} \times \mathbf{R}^{2n} \times X_1.$$

Consequently $\langle L\cdot,\cdot\rangle$ vanishes on $E_{\pm i\mu}$.

REMARK 8.3. *The embedding from $\mathbf{R}^{2n}$ to $\{0\} \times \mathbf{R}^{2n} \times \{0\} \subset X$– an invariant subspace under JL, serves as a similarity transformation between the $2n$-dim Hamiltonian operator $J_{2n}B$ and the restriction of the infinite dimensional one JL. If $i\mu \in \sigma(J_{2n}B)$, then $J_{2n}B$ and JL have exactly the same structures on the subspaces $E_{i\mu}(J_{2n}B)$ and $E_{i\mu}$ of generalized eigenvectors of $i\mu$. However the energy structure is completely destroyed. Namely the $2n$-dim Hamiltonian operator $J_{2n}B$ has a non-trivial energy $\langle B\cdot,\cdot\rangle$ while the energy $\langle L\cdot,\cdot\rangle$ of JL vanishes completely on $\{0\} \times \mathbf{R}^{2n} \times \{0\} \subset X$.*

PROOF. Using the invariance under JL of $\{0\} \times \mathbf{R}^{2n} \times \{0\}$ and $\{0\} \times \mathbf{R}^{2n} \times X_1$, it is easy to compute inductively

$$(JL - i\mu)^k = \begin{pmatrix} (J_{2n}B - i\mu)^k & 0 & 0 \\ A_{21} & (J_{2n}B - i\mu)^k & A_{23} \\ A_{31} & 0 & (A - i\mu)^k \end{pmatrix},$$

where

$$A_{31} = \Sigma_{l=0}^{k-1}(A - i\mu)^l A_1 (J_{2n}B - i\mu)^{k-1-l}.$$

Let $P_{1,2,3}$ denote the projections from X to its components. For any $u = (x_1, x_2, v)^T \in X$, we have

$$\begin{aligned} P_3(JL - i\mu)^k u &= A_{31}x_1 + (A - i\mu)^k v \\ &= A_1(J_{2n}B - i\mu)^{k-1}x_1 + (A - i\mu)\Big((A - i\mu)^{k-1}v \\ &\qquad + \Sigma_{l=1}^{k-1}(A - i\mu)^{l-1}A_1(J_{2n}B - i\mu)^{k-1-l}x_1\Big). \end{aligned}$$

Suppose $P_3(JL - i\mu)^k u = 0$. Since A_1 and $A - i\mu$ are both one-to-one and $R(A_1) \cap R(A - i\mu) = \{0\}$, we obtain

$$(J_{2n}B - i\mu)^{k-1}x_1 = 0,\ (A - i\mu)^{k-1}v + \Sigma_{l=1}^{k-1}(A - i\mu)^{l-1}A_1(J_{2n}B - i\mu)^{k-1-l}x_1 = 0.$$

Let $m \in [0, k-1]$ be the minimal non-negative integer satisfying $(J_{2n}B - i\mu)^m x_1 = 0$. If $m \geq 1$, from the definition of m, the above second equality and the injectivity of $(A - i\mu)^{k-1-m}$ imply

$$\begin{aligned}0 &= (A - i\mu)^m u + \Sigma_{l=k-m}^{k-1}(A - i\mu)^{l+m-k} A_1 (J_{2n}B - i\mu)^{k-1-l} x_1 \\ &= (A - i\mu)\Big((A - i\mu)^{m-1} v + \Sigma_{l=k-m+1}^{k-1}(A - i\mu)^{l+m-k-1} A_1 (J_{2n}B - i\mu)^{k-1-l} x_1\Big) \\ &\quad + A_1 (J_{2n}B - i\mu)^{m-1} x_1.\end{aligned}$$

Again since A_1 and $A - i\mu$ are both one-to-one and $R(A_1) \cap R(A - i\mu) = \{0\}$, we derive $(J_{2n}B - i\mu)^{m-1} x_1 = 0$ which contradicts the definition of m. Therefore $m = 0$, that is, $x_1 = 0$. Due to the injectivity of $(A - i\mu)^k$, it implies $v = 0$ as well.

Suppose $u \in \ker(JL - i\mu)^k$, the above arguments imply $u = (0, x, 0)^T$ and the lemma follows immediately. □

In the rest of this section we will prove that the degeneracy of $\langle L\cdot, \cdot\rangle$ may occur on $E_{i\mu}$ only if $i\mu \in \sigma(JL)$ is not an isolated spectral point.

LEMMA 8.8. *Assume* (**H1-3**) *and $i\mu \in \sigma(JL) \cap i\mathbf{R}$ is isolated in $\sigma(JL)$, then there exist closed subspaces $I^{i\mu}, E_{\&} \subset X$ such that*

(i) *$I^{i\mu}$ and $E_{\&}$ are complexifications of real subspaces of X, namely $u \in I^{i\mu}$ (or $E_{\&}$) if and only if $\overline{u} \in I^{i\mu}$ (or $E_{\&}$). Moreover, they are invariant under JL and*

$$X = I^{i\mu} \oplus E_{\&}, \quad \sigma(JL|_{I^{i\mu}}) = \{\pm i\mu\}, \quad \sigma(JL|_{E_{\&}}) = \sigma(JL)\backslash\{\pm i\mu\}.$$

(ii) $\ker L \subset E_{\&}$ *if $\mu \neq 0$ or* $\ker L \subset I^{i\mu}$ *if $\mu = 0$.*

(iii) $\langle Lu, v\rangle = 0$ *for all $u \in I^{i\mu}$ and $v \in E_{\&}$. Moreover, $\langle L\cdot, \cdot\rangle$ is non-degenerate on quotient spaces* $I^{i\mu}/(\ker L \cap I^{i\mu})$ *and* $E_{\&}/(\ker L \cap E_{\&})$.

PROOF. Let $\Gamma \subset \mathbf{C}\backslash\sigma(JL)$ be a small circle, oriented counterclockwisely, enclosing $i\mu$ but no other elements in $\sigma(JL)$. Define the spectral projection and the eigenspaces

$$P_{i\mu} = \frac{1}{2\pi i}\oint_\Gamma (\lambda - JL)^{-1} d\lambda, \quad E^{i\mu} = P_{i\mu}X.$$

It is standard to verify that $P_{i\mu}$ is a bounded projection on X satisfying

$$JLP_{i\mu} = P_{i\mu}JL; \ \sigma\big((JL)|_{E^{i\mu}}\big) = \{i\mu\}; \ E^{i\mu} \subset D(JL); \ e^{tJL}E^{i\mu} = E^{i\mu}, \ \forall t \in \mathbf{R}.$$

By Lemma 3.6, $-i\mu \in \sigma(JL)$ is also an isolated point of $\sigma(JL)$. Let $P_{-i\mu}$ and $E^{-i\mu}$ be defined similarly. It is standard that $P_{i\mu}P_{-i\mu} = P_{-i\mu}P_{i\mu} = 0$ and thus $P_{i\mu} + P_{-i\mu}$ is also a projection (or $P_{i\mu}$ instead if $\mu = 0$). Define

$$E_{\&} = \ker(P_{i\mu} + P_{-i\mu}), \quad I^{i\mu} = E^{i\mu} + E^{-i\mu}$$

and we have

$$e^{tJL}E_{\&} = E_{\&}, \ \forall t \in \mathbf{R}; \ \sigma\big((JL)|_{E_{\&}}\big) = \sigma(JL)\backslash\{\pm i\mu\}; \ X = I^{i\mu} \oplus E_{\&}. \tag{8.10}$$

Therefore statements (i) and (ii) in the lemma follow from the standard spectral theory. The L-orthogonality between $I^{i\mu}$ and $E_{\&}$ follows from Lemma 6.2 where Ω can be taken as the union of the two small disks centered at $\pm i\mu$.

To complete the proof of the lemma, it suffices to prove the non-degeneracy of $\langle L\cdot, \cdot\rangle$ on $I^{i\mu}/(\ker L \cap I^{i\mu})$ and $E_{\&}/(\ker L \cap E_{\&})$. According to Lemma A.3, $L_{I^{i\mu}}$ and $L_{E_{\&}}$ satisfy (**H2**). Therefore either they are non-degenerate or have non-trivial kernels. Suppose there exists $u \in I^{i\mu}$ such that $\langle Lu, v\rangle = 0$ for all $v \in I^{i\mu}$. From

$X = I^{i\mu} \oplus E_{\&}$ and the L-orthogonality between $I^{i\mu}$ and $E_{\&}$, we obtain $\langle Lu, v\rangle = 0$ for all $v \in X$, which implies $u \in \ker L$. Therefore $\langle L\cdot, \cdot\rangle$ is non-degenerate on $I^{i\mu}/(\ker L \cap I^{i\mu})$. The proof of the non-degeneracy of $\langle L\cdot, \cdot\rangle$ on $E_{\&}/(\ker L \cap E_{\&})$ is similar and thus we complete the proof of the lemma. □

Notice that $I^{i\mu}$ is given not in terms of $E_{i\mu}$, but of $E^{i\mu}$ defined using spectral integrals. In the following we establish the relationship between $I^{i\mu}$ and the subspace $E_{i\mu}$ of generalized eigenvectors.

LEMMA 8.9. *It holds* $I^{i\mu} = E_{i\mu} + E_{-i\mu}$.

PROOF. Let $P^\mu : X \to I^{i\mu}$ be the projection associated to the L-orthogonal decomposition $X = I^{i\mu} \oplus E_{\&}$. Let $J^\mu = P^\mu J (P^\mu)^*$. As $I^{i\mu} \subset D(JL)$, Lemma A.3 implies that $(I^{i\mu}, L_{I^{i\mu}}, J^\mu)$ satisfies assumptions (**H1-3**). The invariance of $I^{i\mu}$ under JL implies $JL|_{I^{i\mu}} = J^\mu L_{I^{i\mu}}$ and $\sigma(J^\mu L_{I^{i\mu}}) = \{\pm i\mu\}$. Since $\pm i\mu \notin \sigma(JL|_{E_{\&}})$, we have $E_{\pm i\mu} \subset I^{i\mu}$.

We apply Theorem 2.1 to JL on $I^{i\mu}$, where there is no hyperbolic subspace, and obtain the decomposition of $I^{i\mu}$ into closed subspaces $I^{i\mu} = \Sigma_{j=0}^4 X_j$, where $X_0 = \ker L$ if $\mu = 0$ or $X_0 = \{0\}$ if $\mu \neq 0$. In this decomposition, $L_{I^{i\mu}}$ and JL take the block forms

$$JL \leftrightarrow \begin{pmatrix} 0 & A_{01} & A_{02} & A_{03} & A_{04} \\ 0 & A_1 & A_{12} & A_{13} & A_{14} \\ 0 & 0 & A_2 & 0 & A_{24} \\ 0 & 0 & 0 & A_3 & A_{34} \\ 0 & 0 & 0 & 0 & A_4 \end{pmatrix}, \quad L_{I^{i\mu}} \leftrightarrow \begin{pmatrix} 0 & 0 & 0 & 0 & 0 \\ 0 & 0 & 0 & 0 & B_{14} \\ 0 & 0 & L_{X_2} & 0 & 0 \\ 0 & 0 & 0 & L_{X_3} & 0 \\ 0 & B_{14}^* & 0 & 0 & 0 \end{pmatrix}.$$

Note $L_{X_3} \geq \delta$ for some $\delta > 0$ and $A_{2,3}$ are anti-self-adjoint with respect to the equivalent inner product $\mp\langle L_{X_{2,3}}\cdot, \cdot\rangle$ with $\sigma(A_{1,2,3,4}) = \{i\mu, -i\mu\}$.

In the case of $\mu = 0$, the anti-self-adjoint operator $A_{2,3}$ must be $A_{2,3} = 0$. Meanwhile all other finite dimensional diagonal blocks are also nilpotent. Therefore it is straightforward to compute that $(JL|_{I^{i\mu}})^k = 0$ for some integer $k > 0$. Therefore $I^{i\mu}$ consists of generalized eigenvectors only and $I^{i\mu} = E_{i\mu}$ in the case of $\mu = 0$.

In the case of $\mu \neq 0$, $X_0 = \{0\}$. Moreover, as A_3 is anti-self-adjoint with respect to the inner product $\langle L_{X_3}\cdot, \cdot\rangle$, we can further decompose X_3 into closed subspaces $X_3 = X_{3+} \oplus X_{3-}$, where $X_{3\pm} = \ker(A_3 \pm i\mu)$, with associated projections $Q_\pm : X_3 \to X_{3\pm}$. Accordingly $A_3 = i\mu Q_+ - i\mu Q_-$, which implies $A_3^2 + \mu^2 = 0$. As $A_{1,2,4}$ are finite dimensional with the only eigenvalues $\pm i\mu$, we obtain that $\left((JL|_{I^{i\mu}})^2 + \mu^2\right)^k = 0$ for some integer $k > 0$. Rewrite it as

$$(JL - i\mu)^k (JL + i\mu)^k = (JL + i\mu)^k (JL - i\mu)^k = 0 \text{ on } I^{i\mu}.$$

Let $X_\pm$ be the invariant eigenspace of $\pm i\mu$ of $JL|_{I^{i\mu}}$ defined via spectral integrals. We have $I^{i\mu} = X_+ \oplus X_-$. As $JL \pm i\mu$ is an isomorphism from $X_\pm$ to itself, we obtain from the above identity that $X_\pm = \ker(JL \mp i\mu)^k$. Therefore $X_\pm$ are the subspaces of generalized eigenvectors of $\pm i\mu$ of JL, that is,

$$I^{i\mu} = \ker(JL - i\mu)^k \oplus \ker(JL + i\mu)^k = E_{i\mu} \oplus E_{-i\mu}.$$

□

Finally, let $E_\# = E_{\&}$ if $\mu = 0$ or $E_\# = E_{-i\mu} \oplus E_{\&}$ if $\mu \neq 0$. In the case of $\mu \neq 0$, Lemmas 3.1, A.2, A.3 and the non-degeneracy of $\langle L\cdot, \cdot\rangle$ on $I^{i\mu}$ imply that

$\langle L\cdot,\cdot\rangle$ is non-degenerate on $E_{i\mu}$ and $E_{\&}/\ker L$. This along with the above lemmas completes the proof of Proposition 2.3.

Based on Proposition 2.3, we are ready to prove Proposition 2.4.

Proof of Proposition 2.4: Let

$$\Lambda = \{0 \neq i\mu \in \sigma(JL) \cap i\mathbf{R} \mid k^{\leq 0}(i\mu) > 0\},$$

which is a finite set according to (2.13) of Theorem 2.3.

Let $i\mu \in \Lambda$. We have that $\langle L\cdot,\cdot\rangle$ is non-degenerate on $E_{i\mu}$ either by our assumption if $i\mu$ is not isolated in $\sigma(JL)$ or by Proposition 2.3 if $i\mu$ is isolated. From Proposition 2.2, we have the L-orthogonal and JL-invariant decomposition $E_{i\mu} = E^1_{i\mu} \oplus E^G_{i\mu}$, where $E^1_{i\mu} \subset \ker(JL - i\mu)$, $\dim E^G_{i\mu} < \infty$, and $\langle L\cdot,\cdot\rangle$ is non-degenerate on both $E^1_{i\mu}$ and $E^G_{i\mu}$. Let $E^{1,-}_{i\mu} \subset E^1_{i\mu}$ be a subspace such that $\dim E^{1,-}_{i\mu} = n^-(L|_{E^-_{i\mu}})$ and $\langle L\cdot,\cdot\rangle$ is negative definite on $E^{1,-}_{i\mu}$. Let

$$E^{finite}_{i\mu} = E^G_{i\mu} \oplus E^{1,-}_{i\mu}, \text{ which satisfies } \dim E^{finite}_{i\mu} < \infty,\ n^-(L|_{E^{finite}_{i\mu}}) = k^{\leq 0}(i\mu).$$

Moreover, $JL(E^{finite}_{i\mu}) = E^{finite}_{i\mu}$ according to its construction.

If $0 \notin \sigma(JL)$, let $E^{finite}_0 = \{0\}$ and we may skip to the next step to define N and M. Otherwise, our assumption and Propositions 2.3, 2.2 imply an L-orthogonal decomposition $E_0 = \ker L \oplus E^1_0 \oplus E^G_0$, where

$$JL(E^1_0) \subset \ker L,\ \dim E^G_0 < \infty,\ JL(E^G_0) \subset E^G_0 \oplus \ker L,$$

and $\langle L\cdot,\cdot\rangle$ is non-degenerate on both E^1_0 and E^G_0. Let $E^{1,-}_0 \subset E^1_0$ be such that $\dim E^{1,-}_0 = n^-(L|_{E^1_0})$ and $\langle L\cdot,\cdot\rangle$ is negative definite on $E^{1,-}_0$. Let $E^{finite}_0 = E^{1,-}_0 \oplus E^G_0$, which satisfies

$$\dim E^{finite}_0 < \infty,\ n^-(L|_{E^{finite}_0}) = k^{\leq 0}_0.$$

Let

$$N = (\oplus_{\mathrm{Re}\,\lambda \neq 0} E_\lambda) \oplus (\oplus_{i\mu \in \Lambda} E^{finite}_{i\mu}) \oplus E^{finite}_0, \quad \tilde{M} = N^{\perp_L} = (N \oplus \ker L)^{\perp_L}.$$

Clearly $\dim N < \infty$, $n^-(L|_N) = n^-(L)$ (due to (2.13) of Theorem 2.3), $\langle L\cdot,\cdot\rangle$ is non-degenerate on N, and $N \oplus \ker L$ is invariant under JL. Therefore $\tilde{M}$ is also invariant under JL, $X = N \oplus \tilde{M}$ (due to Lemma A.2), $\ker L \subset \tilde{M}$, and $n^-(L|_{\tilde{M}}) = 0$. Moreover, N and M are complexifications of real subspaces as E_λ and $E_{\bar{\lambda}}$ have exactly the same structure. Let $M \subset \tilde{M}$ be any closed subspace such that $\tilde{M} = M \oplus \ker L$ and this completes the proof of proposition. □

To end this chapter, we prove the decomposition result Proposition 2.5 for L-self-adjoint operators.

Proof of Proposition 2.5: In order to apply the previous results, which have been given in the framework of real Hilbert spaces, to prove this proposition, we first convert it into a problem on real Hilbert spaces. Recall $(\cdot,\cdot)$ and $\langle\cdot,\cdot\rangle$ denote the complex inner product and the complex duality pair between X^* and X, respectively. Let X_r be the same set as X but equipped with the real inner product $(u,v)_r = \mathrm{Re}(u,v)$. On X_r, the $i-$multiplication $i : X \to X$ becomes a real linear isometry $i_r : X_r \to X_r$ with $i_r^2 = -I$. Let $L_r : X_r \to X_r^*$ be the linear symmetry bounded operator defined as $\langle L_r u, v\rangle_r = \mathrm{Re}\langle Lu, v\rangle$ where $\langle\cdot,\cdot\rangle_r$ denote the real duality pair between X_r^* and X_r. Subsequently the non-degeneracy of L yields the non-degeneracy of L_r. Accordingly A becomes a real linear operator

$A_r : X_r \supset D(A_r) \to X_r$. The linearity of L and A implies that $i_r A_r = A_r i_r$ and $L_r i_r = -i_r^* L_r$. Finally that A is L-self-adjoint is translated to the L_r-self-adjointness of A_r, namely, $L_r A_r = A_r^* L_r$.

Define $J = i_r A_r L_r^{-1} : X_r^* \to X_r$. The L_r-self-adjointness of A_r implies $J^* = -J$ and thus $i_r A_r = J L_r$ with (J, L_r, X_r) satisfying (**H1-3**). It is easy to prove

$$\sigma(A_r) = \sigma(A) \subset \mathbf{R}, \quad \sigma(i_r A_r) = \big(i\sigma(A_r)\big) \cup \big(-i\sigma(A_r)\big),$$

so the nonzero eigenvalues of $i_r A_r$ are isolated and $\ker(i_r A_r) = \ker A$. It is straightforward to deduce the non-degeneracy of $\langle L_r \cdot, \cdot \rangle_r$ on $\ker(i_r A_r)$ from our non-degeneracy assumption of $\langle L\cdot, \cdot\rangle$ on $\ker A$, and thus (**H4**) is satisfied. Thus by Proposition 2.4, there exists a decomposition $X_r = \tilde N \oplus \tilde M$ such that $\tilde N$ and $\tilde M$ are L_r-orthogonal and invariant under $i_r A_r$, $\dim \tilde N < \infty$ and $L_r|_{\tilde M} > 0$, which also implies L is uniformly positive on $\tilde M$. Let

$$N = \tilde N + i_r \tilde N, \quad M = N^{\perp_{L_r}} = \{u \in X_r \mid \langle L_r u, v\rangle_r = 0, \ \forall v \in N\} \subset \tilde M.$$

Clearly $\dim N \le 2\dim \tilde N < \infty$ and $\langle L_r \cdot, \cdot\rangle$ is uniformly positive on M, thus so is $\langle L\cdot, \cdot\rangle$ on M. To complete the proof, we only need to show $N, M \subset X$ are L-orthogonal and invariant under A. We first consider the L-orthogonality which also involves the imaginary part of the quadratic form of L. Suppose there exist $u_1, u_2, \in \tilde N$ and $v \in M$ such that $\langle L(u_1 + iu_2), v\rangle = Re^{i\theta} \ne 0$. It implies $\langle Lu, v\rangle = \langle L_r u, v\rangle_r = R \in \mathbf{R}\backslash\{0\}$, where

$$u = (\cos\theta) u_1 + (\sin\theta) u_2 + i_r\big((\cos\theta) u_2 + (\sin\theta) u_1\big) \in N$$

which is a contradiction to the definitions of N and M and thus they are L-orthogonal. Secondly, $i_r A_r(\tilde N) \subset \tilde N$, $i_r A_r = A_r i_r$, and $i_r^2 = -I$ imply $A_r(i_r \tilde N) \subset \tilde N$ and $A_r \tilde N \subset i_r \tilde N$. Therefore N is invariant under A. It along with the L-self-adjointness of A also implies the invariance of M and the proof of Proposition 2.5 is complete. $\square$

CHAPTER 9

Perturbations

In this chapter, we study the robustness of the spectral properties of the Hamiltonian operator JL under small perturbations preserving Hamiltonian structures. Consider

$$u_t = J_\# L_\# u, \qquad J_\# = J + J_1, \quad L_\# = L + L_1, \quad u \in X.$$

Unless otherwise specified, assumptions (**A1-3**) given in Section 2.5 are assumed throughout this chapter. We first prove

LEMMA 9.1. *Assumptions (**A1-3**) imply that there exists $\epsilon > 0$ depending on J and L such that, if $|L_1| \le \epsilon$, then (**H1-3**) is satisfied by $J_\#$ and $L_\#$ and*

$$\dim \ker L_\# \le \dim \ker L < \infty, \quad D(J_\# L_\#) = D(JL).$$

PROOF. It is obvious that (**H1**) is satisfied by $J_\#$. Let $X_\pm$ be the subspaces provided in (**H2**) satisfied by L. Clearly we still have, for $\epsilon << 1$,

$$\pm \langle L_\# u, u \rangle \ge \delta ||u||^2, \quad \forall u \in X_\pm$$

for some $\delta > 0$ independent of ϵ. Let $X_1 = X_+ \oplus X_-$. Assumption (**H2**) for L implies that$\langle L\cdot, \cdot \rangle$ restricted to X_1 is non-degenerate, i.e.

$$L_{X_1} = i_{X_1}^* L i_{X_1} : X_1 \to X_1^*,$$

defined as in (A.1), is an isomorphism. Therefore

$$L_{\#, X_1} = i_{X_1}^* L_\# i_{X_1} : X_1 \to X_1^*,$$

as a small bounded perturbation of L_{X_1}, is also an isomorphism. Suppose $u = u_0 + u_1 \in \ker L_\#$, where $u_0 \in \ker L$ and $u_1 \in X_1$, then we have

$$0 = L_\# u = L_\# u_1 + L_1 u_0 \implies u_1 = -L_{\#, X_1}^{-1} i_{X_1}^* L_1 u_0,$$

that is,

$$\ker L_\# \subset Y, \text{ where } Y = \text{graph}(S) \text{ and } S = -L_{\#, X_1}^{-1} i_{X_1}^* L_1 : \ker L \to X_1.$$

Moreover, from the (A.2) type identity, it also holds that, for any $v \in \ker L$ and $u_1 \in X_1$,

$$\begin{aligned}\langle L_\#(v + Sv), u_1 \rangle =& \langle L_1 v, u_1 \rangle - \langle L_\# L_{\#, X_1}^{-1} i_{X_1}^* L_1 v, u_1 \rangle \\ =& \langle L_1 v, u_1 \rangle - \langle i_{X_1}^* L_1 v, u_1 \rangle = 0,\end{aligned}$$

that is, Y and X_1 are $L_\#$-orthogonal.

Since $\dim Y = \dim \ker L < \infty$ due to (**A2**), the quadratic form $\langle L_\# \cdot, \cdot \rangle$ restricted to Y leads to a decomposition of Y

$$Y = Y_+ \oplus \ker L_\# \oplus Y_-,$$

where $\pm L_\#$ is positive on $Y_\pm$. Let $X_{\#\pm} = X_\pm \oplus Y_\pm$, then

$$X = X_{\#+} \oplus \ker L_\# \oplus X_{\#-}.$$

Due to the $L_\#$-orthogonality between Y and X_1, it is easy to derive that $\pm\langle L_\#\cdot, \cdot\rangle$ are positive definite on $X_{\#\pm}$. Therefore (**H2**) is satisfied.

Finally we prove (**H3**). Suppose $\gamma \in X^*$ and $\langle \gamma, u\rangle = 0$ for all $u \in X_{\#+} \oplus X_{\#-} \supset X_+ \oplus X_-$. From (**A1**) which requires that (**H1-3**) being satisfied by J and L, we have $\gamma \in D(J) = D(J_\#)$ as J_1 is assumed to be bounded. $\square$

Much as in Remark 2.2 by composing with the Riesz representation, we may treat $L_\#$ as a bounded symmetric operator on X and then apply its spectral decomposition, a decomposition satisfying (**H2**) can be obtained much more easily. However, that decomposition may not satisfy (**H3**).

In Section 9.1, we will obtain the persistence of exponential trichotomy of the perturbed system. In Section 9.2, we will focus on purely imaginary spectral points of $\sigma(JL)$ and the possibility of bifurcation of unstable eigenvalues of $J_\# L_\#$. To start, following the standard procedure we show that assumption (**A3**) implies the convergence of the resolvents. Recall $||\cdot||_G$ denote the graph norm on $D(JL)$ and $|\cdot|_G$ the corresponding operator norm.

LEMMA 9.2. *Let $K \subset \mathbf{C}\backslash\sigma(JL)$ be compact, then there exist $C, \epsilon > 0$ depending on K, J, and L, such that, for any $\lambda \in K$ and*

$$|J_1|, |JL_1|_G \le \epsilon, \quad |L_1| \le 1,$$

it holds that the densely defined closed operator $\lambda - J_\# L_\# : D(JL) \to X$ has a bounded inverse and

$$|(\lambda - J_\# L_\#)^{-1} - (\lambda - JL)^{-1}| \le C(|J_1| + |JL_1|_G).$$

PROOF. It is straightforward to compute

$$\lambda - J_\# L_\# = \big(I - (JL_1 + J_1 L_\#)(\lambda - JL)^{-1}\big)(\lambda - JL).$$

According to assumption (**A3**), $JL_1(\lambda - JL)^{-1}$ is a closed operator with the domain X. The closed graph theorem implies that it is actually bounded with

$$\begin{aligned} |JL_1(\lambda - JL)^{-1}| \le& |JL_1|_G\big(|(\lambda - JL)^{-1}| + |JL(\lambda - JL)^{-1}|\big) \\ \le& |JL_1|_G\big(1 + (1 + |\lambda|)|(\lambda - JL)^{-1}|\big) \end{aligned}$$

where $JL = \lambda - (\lambda - JL)$ was used in the last step. The conclusion of the lemma follows from this along with the boundedness of J_1, L, and L_1. $\square$

9.1. Persistent exponential trichotomy and stability: Theorem 2.4 and Proposition 2.9

In this section, our main task is to prove Theorem 2.4 as well as Proposition 2.9. With the help of Lemmas 9.2 and 6.2, we are able to prove most of Theorem 2.4 and Proposition 2.9 by standard arguments in the spectral theory. However, proving (2.26) requires more elaborated arguments as one of the perturbation term JL_1 is not necessarily a small bounded operator.

Proof of Theorem 2.4 except (2.26). Adopt the notation used in (2.24)

$$|J_1| + |L_1| + |JL_1|_G \le \epsilon. \tag{9.1}$$

Let

$$\sigma_{u,s} = \{\lambda \in \sigma(JL) \mid \pm \mathrm{Re}\,\lambda > 0\} \subset \sigma(JL)$$

and $\Omega_u \subset \mathbf{C}$ be open and bounded with smooth boundary $\Gamma_u = \partial\Omega_u \subset \mathbf{C}\backslash\sigma(JL)$ such that $\sigma(JL) \cap \Omega_u = \sigma_u$. According to Lemma 3.6, σ_s is symmetric to σ_u about $i\mathbf{R}$ and thus we let Ω_s be the domain symmetric to Ω_u and $\Gamma_s = \partial\Omega_s$. For small ϵ, Lemma 9.2 allows us to define the following objects via standard contour integrals

$$\tilde{P}_{\#}^{u,s} = \frac{1}{2\pi i}\oint_{\Gamma_{u,s}} (z - J_{\#}L_{\#})^{-1} dz, \quad E_{\#}^{u,s} = \tilde{P}_{\#}^{u,s} X,$$

$$A_{\#}^{u,s} = (J_{\#}L_{\#})|_{E_{\#}^{u,s}} = \frac{1}{2\pi i}\oint_{\Gamma_{u,s}} z(z - J_{\#}L_{\#})^{-1} dz.$$

Let

$$\tilde{P}_{\#}^{c} = I - \tilde{P}_{\#}^{u} - \tilde{P}_{\#}^{s}, \quad E_{\#}^{c} = \tilde{P}_{\#}^{c} X, \quad A_{\#}^{c} = (J_{\#}L_{\#})|_{E_{\#}^{c}},$$

and $\tilde{P}^{u,s}$, $E^{u,s,c}$, $A^{u,s,c}$ denote the corresponding unperturbed objects.

From the standard spectral theory, subspaces $E_{\#}^{u,s,c}$ are invariant under $J_{\#}L_{\#}$. Therefore $A_{\#}^{u,s,c}$ are operators on $E_{\#}^{u,s,c}$ with $\sigma(A_{\#}^{u,s}) \subset \Omega_{u,s}$ and $\sigma(A_{\#}^{c}) \subset \big(\mathbf{C}\backslash(\Omega_u \cup \Omega_s)\big)$. Lemma 9.2 implies

$$|\tilde{P}_{\#}^{u,s} - \tilde{P}^{u,s}| \le C\epsilon,$$

and thus $E_{\#}^{c}$ is $O(\epsilon)$ close to E^c, too. Along with the non-degeneracy of $\langle L\cdot,\cdot\rangle$ on $E^u \oplus E^s$ and $|L_1| \le \epsilon$, above implies the non-degeneracy of $\langle L_{\#}\cdot,\cdot\rangle$ on $E_{\#}^{u} \oplus E_{\#}^{s}$. Therefore we obtain from $X = E_{\#}^{u} \oplus E_{\#}^{s} \oplus E_{\#}^{c}$ and Lemma 6.2

$$E_{\#}^{c} = \{u \in X \mid \langle L_{\#}u, v\rangle = 0, \ \forall u \in E_{\#}^{u} \oplus E_{\#}^{s}\}.$$

As $O(\epsilon)$ perturbations, it is clear that subspaces $E_{\#}^{u,s,c}$ can be written as graphs of $O(\epsilon)$ bounded operators $S_{\#}^{u,s,c}$ in the coordinate frame $X = E^u \oplus E^s \oplus E^c$. Moreover, from the above integral forms, $A_{\#}^{u,s}$ are only $O(\epsilon)$ bounded perturbations to JL on finite dimensional subspaces $E_{\#}^{u,s}$ which are $O(\epsilon)$ perturbations to $E^{u,s}$, and thus inequality (2.25) follows as well. Since the subspace $E_{\#}^{u} \oplus E_{\#}^{s}$, invariant under $J_{\#}L_{\#}$, is finite dimensional, the vanishness of $\langle L_{\#}\cdot,\cdot\rangle$ on $E_{\#}^{u,s}$ follows from Lemma 3.3. Through this point we complete the proof of parts (a) and (b), except (2.26), of Theorem 2.4.

Suppose, as in part (c) in Theorem 2.4, there exists $\delta > 0$ such that $\langle Lu, u\rangle \ge \delta||u||^2$ for all $u \in E^c$. Since L and L_1 are bounded and $E_{\#}^{c}$ is $O(\epsilon)$ perturbation of E^c, we have $\langle L_{\#}u, u\rangle \ge \frac{\delta}{2}||u||^2$ for all $u \in E_{\#}^{c}$. Therefore the conservation of $\langle L_{\#}\cdot,\cdot\rangle$ by $e^{tJ_{\#}L_{\#}}$ and the invariance of $E_{\#}^{c}$ under $e^{tJ_{\#}L_{\#}}$ imply the boundedness of $e^{tJ_{\#}L_{\#}}|_{E_{\#}^{c}}$ uniformly in $t \in \mathbf{R}$ which proves part (b) of Theorem 2.4. □

To complete the proof of Theorem 2.4, we shall prove the weak exponential growth estimate (2.26) in the perturbed center subspace $E_{\#}^{c}$, which involves much more than simple applications of the standard operator calculus and the conservation of energy. We first consider a special case where JL has no hyperbolic directions.

LEMMA 9.3. *Assume $E^c = X$, then (2.26) holds for some $C, \epsilon_0 > 0$ depending on J, L.*

PROOF. From the construction of $E_{\#}^{u,s,c}$ and the additional assumption $E^c = X$, it is clear $E_{\#}^c = X$ and (2.26) is reduced to

$$|e^{tJ_{\#}L_{\#}}| \le Ce^{C\epsilon|t|}, \quad \forall t \in \mathbf{R}, \quad \text{where } \epsilon \triangleq |J_1| + |L_1| + |JL_1|_G.$$

Since

$$J_{\#}L_{\#} = JL_{\#} + J_1(L + L_1)$$

and J_1, L, and L_1 are bounded with $|J_1| \le \epsilon$, we have

$$|J_{\#}L_{\#} - JL_{\#}| \le C\epsilon. \tag{9.2}$$

Let $X = \oplus_{j=0}^4 X_j$ be the decomposition, associated with projections P_j, given by Theorem 2.1 for J and L, where $X_0 = \ker L$ and $X_5 = X_6 = \{0\}$ due to the assumption $E^c = X$. Much as in (A.1), let

$$L_{1,jk} = i_j^* L_1 i_k : X_k \to X_j^*, \quad j, k = 0, \ldots, 4,$$

which satisfy $L_{1,jk} = L_{1,kj}^*$ and

$$L_1 = \Sigma_{j,k=0}^4 P_j^* L_{1,jk} P_k, \qquad \langle L_{1,jk} u, v\rangle = \langle L_1 u, v\rangle, \quad \forall u \in X_k,\ v \in X_j.$$

Let $J_{jk} = P_j J P_k^*$ be the blocks of J associated to this decomposition, which have the forms given in Corollary 2.1 and satisfy

$$J = \Sigma_{j,k=0}^4 i_{X_j} J_{jk} i_{X_k}^*, \quad |J - i_{X_3} J_{33} i_{X_3}^*| \le C.$$

We write

$$\begin{aligned} JL_{\#} =& JL + JL_1P_3 + \Sigma_{k\in\{0,1,2,4\}} JL_1P_k \\ =& JL + (J - i_{X_3}J_{33}i_{X_3}^*)L_1P_3 \\ &+ i_{X_3}J_{33}i_{X_3}^* \Sigma_{j,k=0}^4 P_j^* L_{1,jk}P_kP_3 + \Sigma_{k\in\{0,1,2,4\}} JL_1P_k \\ =& JL + i_{X_3}J_{33}L_{1,33}P_3 + (J - i_{X_3}J_{33}i_{X_3}^*)L_1P_3 + \Sigma_{k\in\{0,1,2,4\}} JL_1P_k, \end{aligned}$$

where $P_{X_j} i_{X_k} = \delta_{jk} I_{X_k}$ is used. Since, for $k \ne 3$, $X_k \subset D(JL) \subset D(JL_1)$, we have that JL_1P_k is a bounded operator with the norm bounded in terms of $|JL_1|_G$, $|P_k|$, and $|JLP_k|$. Along with the boundedness of $J - i_{X_3}J_{33}i_{X_3}^*$, we obtain

$$|JL_{\#} - (JL + i_{X_3}J_{33}L_{1,33}P_3)| < C\epsilon. \tag{9.3}$$

From Theorem 2.1 we have

$$JL + i_{X_3}J_{33}L_{1,33}P_3 \longleftrightarrow \begin{pmatrix} 0 & A_{01} & A_{02} & A_{03} & A_{04} \\ 0 & A_1 & A_{12} & A_{13} & A_{14} \\ 0 & 0 & A_2 & 0 & A_{24} \\ 0 & 0 & 0 & A_3 + J_{33}L_{1,33} & A_{34} \\ 0 & 0 & 0 & 0 & A_4 \end{pmatrix}.$$

Note all blocks of $JL + i_{X_3}J_{33}L_{1,33}P_3$ are identical to those of JL except its $(4,4)$-block

$$A_3 + J_{33}L_{1,33} = J_{33}(L_{X_3} + L_{1,33}).$$

Since $\langle L_{X_3}\cdot, \cdot\rangle$ is uniformly positive on X_3, so is $\langle (L_{X_3} + L_{1,33})\cdot, \cdot\rangle$. Therefore the group $e^{t(A_3 + J_{33}L_{1,33})}$, conserving $\langle (L_{X_3} + L_{1,33})\cdot, \cdot\rangle$, satisfies

$$|e^{t(A_3+J_{33}L_{1,33})}| \le C, \quad \forall t \in \mathbf{R}.$$

By using the upper triangular form of JL, this inequality, assumption $E^c = X$ that JL has no hyperbolic eigenvalues, and the finite dimensionality $\dim X_1 = \dim X_4 = n^-(L) - \dim X_2$, it is easy to prove

$$|e^{t(JL+i_{X_3}J_{33}L_{1,33}P_3)}| \leq C(1+|t|^{1+2n^-(L)}), \quad \forall t \in \mathbf{R}.$$

Along with (9.2), (9.3), and the above estimate, from the following lemma we obtain (2.26) assuming $E^c = X$. $\square$

LEMMA 9.4. *Let $\mathbf{X}$ be a Banach space space, $\omega \in \mathbf{R}$, $C_0, k \geq 0$, and $\mathbf{A} : D(\mathbf{A}) \to X$ the generator of a C^0 semigroup on $\mathbf{X}$, such that*

$$|e^{t\mathbf{A}}| \leq C_0(1+t^k)e^{\omega t}, \quad \forall t > 0.$$

Suppose $\mathbf{A}_1 \in L(\mathbf{X})$ and $|\mathbf{A}_1| \leq \epsilon \in (0,1]$, then there exists $C > 0$ depending only on k and C_0 such that

$$|e^{t(\mathbf{A}+\mathbf{A}_1)}| \leq C\epsilon^{-\frac{k}{k+1}}e^{(\omega+C\epsilon^{\frac{1}{k+1}})t}, \quad \forall t > 0.$$

PROOF. Without loss of generality, we may assume $\omega = 0$ (otherwise $\mathbf{A}$ can be replaced by $\mathbf{A}-\omega$). Since $t^k e^{-t}$ is bounded for $t > 0$, there exists $C > 0$ depending only on k such that $(1+t^k) \leq C\epsilon^{-\frac{k}{k+1}}e^{\epsilon^{\frac{1}{k+1}}t}$ for all $t > 0$. From the variation of parameter formula we have

$$\begin{aligned}|e^{t(\mathbf{A}+\mathbf{A}_1)}| =& \left|e^{t\mathbf{A}} + \int_0^t e^{(t-\tau)\mathbf{A}}\mathbf{A}_1 e^{\tau(\mathbf{A}+\mathbf{A}_1)}d\tau\right| \\ \leq& C\epsilon^{-\frac{k}{k+1}}e^{\epsilon^{\frac{1}{k+1}}t} + C\epsilon^{\frac{1}{k+1}}\int_0^t e^{\epsilon^{\frac{1}{k+1}}(t-\tau)}\left|e^{\tau(\mathbf{A}+\mathbf{A}_1)}\right|d\tau,\end{aligned}$$

and the desired estimate follows from the Gronwall inequality. $\square$

By using the invariance of $E^c_\#$ under $J_\# L_\#$, in the following we convert $e^{tJ_\# L_\#}|_{E^c_\#}$ to a flow $e^{t\tilde{J}_\# \tilde{L}_\#}$ on E^c via a similarity transformation and then apply Lemma 9.3 to obtain (2.26).

Proof of (2.26) in Theorem 2.4. In the general case, let $E^{u,s,c}_\#$ be the invariant unstable/stable/center subspaces and $P^{u,s,c}_\#$ be the projections associated to the decomposition $X = E^u_\# \oplus E^s_\# \oplus E^c_\#$. We also adopt the notations $E^{us}_\# = E^u_\# \oplus E^s_\#$ and $P^{us}_\# = P^u_\# + P^s_\#$. Correspondingly, let $E^{u,s,c,us}$ and $P^{u,s,c,us}$ denoted the unperturbed invariant subspaces and projections. Recall $E^c_\#$ can be written as the graph of a bounded operator $S^c_\# : E^c \to E^{us}$ with $|S^c_\#| = O(\epsilon)$. Let $\tilde{S}^c_\# = i_{E^c} + S^c_\# : E^c \to E^c_\# \subset X$ so that $E^c_\# = \tilde{S}^c_\#(X^c)$. Clearly $P^c i_{E^c_\#} = (\tilde{S}^c_\#)^{-1}$.

Let

$$J^c = P^c J(P^c)^*, \quad J^c_\# = P^c_\# J_\#(P^c_\#)^*, \quad L^c_\# = i^*_{E^c_\#}L_\# i_{E^c_\#}, \quad L^c = i^*_{E^c}Li_{E^c}.$$

From the invariance of $E^c_\#$ under $J_\# L_\#$, the $L_\#$-orthogonality between $E^c_\#$ and $E^{us}_\#$, and Lemma A.3 applied to the decomposition $X = E^c_\# \oplus E^{us}_\#$, we have

$$J^c_\# L^c_\# = J_\# L_\#|_{E^c_\#}, \quad e^{t(J_\# L_\#)}|_{E^c_\#} = e^{tJ^c_\# L^c_\#},$$

and the combination $(E^c_\#, J^c_\#, L^c_\#)$ satisfies (H1-3). Using the mapping $\tilde{S}^c_\#$, we may just consider its conjugate flow on E^c

$$P^c i_{E^c_\#}e^{tJ^c_\# L^c_\#}\tilde{S}^c_\#, \quad \text{with the generator } P^c i_{E^c_\#}J^c_\# L^c_\# \tilde{S}^c_\# \text{ on } E^c.$$

Let

$$\tilde{L}_\# = (\tilde{S}^c_\#)^* L^c_\# \tilde{S}^c_\# = (\tilde{S}^c_\#)^* i^*_{E^c_\#} L_\# i_{E^c_\#} \tilde{S}^c_\# : E^c \to (E^c)^*,$$

and

$$\tilde{J}_\# = P^c i_{E^c_\#} J^c_\# (P^c i_{E^c_\#})^* = P^c P^c_\# J_\# (P^c P^c_\#)^* : (E^c)^* \supset D(\tilde{J}_\#) \to E^c.$$

Clearly

$$\tilde{J}_\# \tilde{L}_\# = P^c i_{E^c_\#} J^c_\# L^c_\# \tilde{S}^c_\# = P^c i_{E^c_\#} J_\# L_\# \tilde{S}^c_\#. \tag{9.4}$$

Since $|P^c i_{E^c_\#}|, |\tilde{S}^c_\#| \le 2$, in order to prove (2.26), it suffices to prove on E^c

$$|e^{\tilde{J}_\# \tilde{L}_\#}| \le C\epsilon^{\frac{1}{2(1+n^-(L)-\dim E^u)}-1} e^{C\epsilon^{\frac{1}{2(1+n^-(L)-\dim E^u)}}|t|}, \ \forall t \in \mathbf{R} \tag{9.5}$$

for some C depending only on J and L. Our strategy is to verify that $(E^c, \tilde{J}_\#, \tilde{L}_\#)$ as a perturbation to (E^c, J^c, L^c) satisfies (**A1-3**) and then apply Lemma 9.3.

When $\epsilon = 0$, Lemma A.3 ensures that the unperturbed

$$(E^c,\ \tilde{J}_\# = J^c,\ \tilde{L}_\# = L^c = i^*_{E^c} L i_{E^c})$$

satisfies (**H1-3**). Moreover, since $\langle L\cdot, \cdot\rangle$ is non-degenerate on E^{us}, we have

$$\dim\ker L^c = \dim\ker(i^*_{E^c} L i_{E^c}) = \dim\ker L < \infty$$

due to the L-orthogonality between E^c and E^{us} and thus (**A2**) is satisfied by $\tilde{L}_\#$ for $\epsilon = 0$. From the definitions, $\tilde{J}_\# - J^c$ is clearly anti-symmetric. We will show that it is also bounded. Using the fact $I - P^c_\# = P^{us}_\#$, one may compute

$$\begin{aligned}\tilde{J}_\# - J^c =& - P^c P^{us}_\# J_\# (P^c P^c_\#)^* - P^c J_\# (P^c P^{us}_\#)^* + P^c J_1 (P^c)^* \\ =& - P^c P^{us}_\# P^{us}_\# J_\# (P^c P^c_\#)^* - P^c J_\# (P^{us}_\#)^* (P^c P^{us}_\#)^* + P^c J_1 (P^c)^*.\end{aligned}$$

Due to the $L_\#$-orthogonality between $E^{us}_\#$ and $E^c_\#$ and the non-degeneracy of $\langle L_\# \cdot, \cdot\rangle$ on $E^{us}_\#$, it is straightforward to obtain that $L_\#$ is an isomorphism from $E^{us}_\#$ to $R\big((P^{us}_\#)^*\big) = (P^{us}_\#)^*(E^{us}_\#)^*$. Since $E^{us}_\# \subset D(J_\# L_\#)$, we have $R\big((P^{us}_\#)^*\big) \subset D(J_\#)$ and thus $J_\#|_{R\left((P^{us}_\#)^*\right)}$ is a bounded operator. To estimate its norm, we use the relationship

$$J_\#|_{R\left((P^{us}_\#)^*\right)} = (J_\# L_\# |_{E^{us}_\#})(L_\# |_{E^{us}_\#})^{-1} = (A^u_\# \oplus A^s_\#)(L_\# |_{E^{us}_\#})^{-1}.$$

Recall $E^{us}_\#$ is $O(\epsilon)$ perturbation to E^{us} and $L_\#$ is $O(\epsilon)$ to L. Moreover, the spectral integral representations of $A^{u,s}_\#$ yield that they are $O(\epsilon)$ perturbation to $JL|_{E^{us}}$. Therefore we obtain that

$$|J_\#|_{R\left((P^{us}_\#)^*\right)}| \le C \Longrightarrow |P^{us}_\# J_\#| = |J_\# (P^{us}_\#)^*| \le C$$

for some $C > 0$ depending on J and L. Since $|P^c P^{us}_\#| \le C\epsilon$, we have

$$|\tilde{J}_{1,\#}| \le C\epsilon, \quad \text{where } \tilde{J}_{1,\#} \triangleq \tilde{J}_\# - J^c.$$

From the definition of $\tilde{L}_\#$, it is easy to obtain

$$\tilde{L}_{1,\#} = \tilde{L}^*_{1,\#}, \quad |\tilde{L}_{1,\#}| \le C\epsilon, \quad \text{where } \tilde{L}_{1,\#} \triangleq \tilde{L}_\# - L^c.$$

Therefore we finish verifying (**A1**) for $(E^c, \tilde{J}_\#, \tilde{L}_\#)$.

We proceed to verify (**A3**). From Lemma 9.2, we have $D(J_\# L_\#) = D(JL)$. Since $E^c_\# = \tilde{S}^c_\#(E^c)$ is the graph of $S^c_\# : E^c \to E^{us}$ and $E^{us} \subset D(JL) = D(J_\# L_\#)$, we obtain

$$D(J_\# L_\#) \cap E^c_\# = \tilde{S}^c_\#\big(E^c \cap D(JL)\big).$$

From the boundedness of $\tilde{J}_{1,\#}$ and (9.4), we further obtain

$$D(J^c \tilde{L}_\#) = D(\tilde{J}_\# \tilde{L}_\#) = E^c \cap D(JL) = D(J^c L^c)$$

which along with $\tilde{L}_\# = L^c + \tilde{L}_{1,\#}$ obviously implies $D(J^c L^c) \subset D(J^c \tilde{L}_{1,\#})$.

In the next we estimate the graph norm of $J^c \tilde{L}_{1,\#}$, like the one defined in (2.23), on the domain $D(J^c \tilde{L}_\#) = E^c \cap D(JL)$. From (9.4) one may compute that, when restricted on $E^c \cap D(JL)$,

$$\tilde{J}_\# \tilde{L}_\# - J^c L^c = -P^{us} i_{E^c_\#} J_\# L_\# \tilde{S}^c_\# + J_\# L_\# S^c_\# + J_\# L_\# - JL. \tag{9.6}$$

We shall use

$$J_\# L_\# = JL + JL_1 + J_1 L_\# \tag{9.7}$$

to estimate the three terms in (9.6). In fact, for any $v \in D(JL)$,

$$||J_\# L_\# v - JLv|| \le |JL_1|_G ||v||_G + |J_1 L_\#| ||v|| \le C\epsilon ||v||_G$$

for some $C > 0$ depending on J and L. A combination of this inequality with (9.6) and the fact $|P^{us} i_{E^c_\#}| \le C\epsilon$ implies, for any $u \in E^c \cap D(JL)$,

$$||(\tilde{J}_\# \tilde{L}_\# - J^c L^c)u|| \le C\big(\epsilon ||\tilde{S}^c_\# u||_G + ||S^c_\# u||_G + \epsilon ||u||_G\big) \le C\epsilon ||u||_G,$$

where we used the fact that JL is bounded on $E^{us} \subset D(JL)$. Since

$$J^c \tilde{L}_{1,\#} = \tilde{J}_\# \tilde{L}_\# - J^c L^c - \tilde{J}_{1,\#} \tilde{L}_\#$$

with $|\tilde{J}_{1,\#}| \le C\epsilon$, the above inequality implies

$$||J^c \tilde{L}_{1,\#} u|| \le C\epsilon ||u||_G, \quad \forall u \in E^c \cap D(JL)$$

for some $C > 0$ depending on J and L. The above estimates allow us to apply Lemma 9.3 to obtain (9.5) for $e^{t\tilde{J}_\# \tilde{L}_\#}$ on E^c, which in turn implies (2.26) for $e^{tJ_\# L_\#}$ on $E^c_\#$. □

To complete this section we present

Proof of Proposition 2.9. Adopt the notations used in (2.24)-(9.1), and let

$$\epsilon \triangleq |J_1| + |L_1| + |JL_1|_G.$$

Let $\Omega \subset \mathbf{C}$ be an open domain with the compact closure and smooth boundary $\Gamma \subset \mathbf{C} \backslash i\mathbf{R}$ such that $\Omega \cap \sigma(JL) = \sigma_2$. For small ϵ, Lemma 9.2 allows us to define the following objects via standard contour integrals

$$P_\# = \frac{1}{2\pi i} \oint_\Gamma (z - J_\# L_\#)^{-1} dz, \quad X_2^\# = P_\# X \subset D(J_\# L_\#), \quad X_1^\# = (I - P_\#)X$$

$$A^\#_{1,2} = (J_\# L_\#)|_{X_{1,2}} = \frac{1}{2\pi i} \oint_\Gamma z(z - J_\# L_\#)^{-1} dz.$$

Let P, $X_{1,2}$, and $A_{1,2}$ denote the corresponding unperturbed objects.

From the standard spectral theory, the decomposition $X = X_1^\# \oplus X_2^\#$ is invariant under $J_\# L_\#$ and thus $A^\#_{1,2}$ are operators on $X^\#_{1,2}$ with $\sigma(A_2^\#) \subset \Omega$. (In fact $\sigma(A_{1,2}) = \sigma_{1,2}$.) Since $\sigma(J_\# L_\#) = \sigma(A_1^\#) \cup \sigma(A_2^\#)$, we only need to prove that $\sigma(A_1^\#) \subset i\mathbf{R}$.

Since the decomposition $X = X_1 \oplus X_2$ is L-orthogonal and $X_2 \subset D(JL)$, Lemma A.3 implies that $(X_1, J_{X_1} = (I-P)J(I-P)^*, L_{X_1})$ satisfies (**H1-3**). Therefore the index theorem Theorem 2.3 applies to L_{X_1} and $JL|_{X_1}$ which along with the second assumption of Proposition 2.9 implies that $n^-(L_{X_1}) = 0$. As L_{X_1} satisfies (**H2**), we obtain that L_{X_1} is positive definite. Lemma 9.2 implies

$$|P_\# - P| \leq C\epsilon.$$

and thus $X_1^\#$ is $O(\epsilon)$ close to X_1. Namely $X_1^\#$ can be written as the graph of an $O(\epsilon)$ order bounded operator $S_\# : X_1 \to X_2$. It immediately implies that $L_{X_1^\#}$ is uniformly positive on $X_1^\#$ and the proposition follows from the invariance of $X_1^\#$ under $J_\# L_\#$. $\square$

9.2. Perturbations of purely imaginary spectrum and bifurcation to unstable eigenvalues

In this section, we consider $\sigma(J_\# L_\#)$ near some $i\mu \in \sigma(JL) \cap i\mathbf{R}$ and prove Theorems 2.5 and 2.6.

'Structurally stable' cases. We still adopt the notation used in (2.24) and let

$$\epsilon \triangleq |J_1| + |L_1| + |JL_1|_G. \tag{9.8}$$

Case 1: $i\mu \in \sigma(JL) \cap i\mathbf{R}$ is isolated with $\langle L\cdot, \cdot\rangle$ sign definite on $E_{i\mu}$.

Suppose $\delta > 0$ and $\langle Lu, u\rangle \geq \delta ||u||^2$, for all $u \in E_{i\mu}$ (the opposite case $\langle L\cdot, \cdot\rangle \leq -\delta < 0$ on $E_{i\mu}$ is similar). Since $i\mu$ is assumed to be isolated in $\sigma(JL)$, there exists $\alpha > 0$ such that the closed disk $\overline{B(i\mu, \alpha)} \cap \sigma(JL) = \{i\mu\}$. Let $\Gamma = \partial B(i\mu, \alpha)$ and $\Gamma \cap \sigma(J_\# L_\#) = \emptyset$ for small ϵ due to Lemma 9.2. Define

$$\tilde{P}_\# = \frac{1}{2\pi i} \oint_\Gamma (z - J_\# L_\#)^{-1} dz, \quad \tilde{E}_\# = \tilde{P}_\# X.$$

From the standard spectral theory, $\tilde{E}_\#$ is invariant under $J_\# L_\#$ and

$$\sigma(J_\# L_\#) \cap \overline{B(i\mu, \alpha)} = \sigma(J_\# L_\# |_{\tilde{E}_\#}).$$

Lemma 9.2, the isolation of $i\mu$, and Proposition 2.3 imply that $\tilde{E}_\#$ is $O(\epsilon)$ close to $E_{i\mu}$. The positive definiteness assumption of $\langle L\cdot, \cdot\rangle$ on $E_{i\mu}$ and the boundedness of L and L_1 imply that $\langle L_\# \cdot, \cdot\rangle$ is also positive definite on $\tilde{E}_\#$. The stability – both forward and backward in time – of $e^{tJ_\# L_\#}$ on $\tilde{E}_\#$, due to the conservation of energy, implies

$$\sigma(J_\# L_\#) \cap \overline{B(i\mu, \alpha)} = \sigma(J_\# L_\# |_{\tilde{E}_\#}) \subset i\mathbf{R}.$$

Case 2: $i\mu \in \sigma(JL) \cap i\mathbf{R}$ and $\langle L\cdot, \cdot\rangle$ is positive definite on $E_{i\mu}$.

Then

$$E_{i\mu} = \{0\} \text{ or } \langle Lu, u\rangle \geq \delta ||u||^2, \ \forall u \in E_{i\mu}, \tag{9.9}$$

for some $\delta > 0$.

REMARK 9.1. *In this case, besides the possibility of an isolated eigenvalue $i\mu \in \sigma(JL)$ with L positive definite on $E_{i\mu}$, we are mainly concerned with the scenario that $i\mu$ is embedded in the continuous spectrum, wether an eigenvalue or not, but without any eigenvector in a non-positive direction of L. Our conclusion is that, under small perturbations, no hyperbolic eigenvalues (i.e. away from imaginary axis) may bifurcate from $i\mu$.*

We argue by contradiction for Case 2. Suppose Theorem 2.5 does not hold in this case, then there exist a sequence

$$J_{\#n} = J + \tilde{J}_n, \; L_{\#n} = L + \tilde{L}_n, \quad n = 1, 2, \dots,$$

satisfying (**A1-3**) for each n such that

$$\exists \lambda_n \in \sigma(J_{\#n} L_{\#n}) \backslash i\mathbf{R}; \;\; \epsilon_n \triangleq |\tilde{J}_n| + |\tilde{L}_n| + |J\tilde{L}_n|_G \to 0; \;\; \delta_n \triangleq |\lambda_n - i\mu| \to 0.$$

Since not in $i\mathbf{R}$, λ_n must be eigenvalues. Let

$$u_n \in X, \quad J_{\#n} L_{\#n} u_n = \lambda_n u_n, \quad ||u_n|| = 1.$$

Using the graph norm of JL_1, one may estimate

$$||J\tilde{L}_n u_n|| \le |J\tilde{L}_n|_G (1 + ||JLu_n||) \le |J\tilde{L}_n|_G \big(1 + |\lambda_n| + ||JLu_n - \lambda_n u_n||\big)$$

and

$$||JLu_n - \lambda_n u_n|| = ||JLu_n - J_{\#n} L_{\#n} u_n|| \le ||\tilde{J}_n L_{\#n} u_n|| + ||J\tilde{L}_n u_n||.$$

Therefore we obtain

$$||JLu_n - \lambda_n u_n|| \le C\epsilon_n, \;\; ||JLu_n - i\mu u_n|| \le (C\epsilon_n + \delta_n) \tag{9.10}$$

for some $C > 0$ depending on $|L|$ and μ.

Let $X = \oplus_{j=0}^{6}$ be the decomposition given by Theorem 2.1 for (L, J), with $X_0 = \ker L$, P_j be the associated projections, and $u_{n,j} = P_j u_n$. Let A_j and A_{jk} denote the blocks of JL in this decomposition as given in Theorem 2.1. From the commutativity between JL and $P_{5,6}$, we obtain from (9.10)

$$||A_5 u_{n,5} - i\mu u_{n,5}|| + ||A_6 u_{n,6} - i\mu u_{n,6}|| \le C(\epsilon_n + \delta_n).$$

Since $\sigma(A_{5,6}) \cap i\mathbf{R} = \emptyset$, we have

$$||u_{n,5}|| + ||u_{n,6}|| \le C(\epsilon_n + \delta_n). \tag{9.11}$$

From Lemma 3.3, we have $\langle L_{\#n} u_n, u_n \rangle = 0$. Along with (9.11) this implies that

$$|2\langle Lu_{n,1}, u_{n,4}\rangle + \langle L_2 u_{n,2}, u_{n,2}\rangle + \langle L_3 u_{n,3}, u_{n,3}\rangle| \le C(\epsilon_n + \delta_n). \tag{9.12}$$

Applying P_j, $j = 0, \dots, 4$ to (9.10) and using Theorem 2.1, we have

$$\begin{aligned}
&||A_4 u_{n,4} - i\mu u_{n,4}|| \le C(\epsilon_n + \delta_n); && (9.13)\\
&||A_3 u_{n,3} + A_{34} u_{n,4} - i\mu u_{n,3}|| \le C(\epsilon_n + \delta_n);\\
&||A_2 u_{n,2} + A_{24} u_{n,4} - i\mu u_{n,2}|| \le C(\epsilon_n + \delta_n);\\
&||A_1 u_{n,1} + A_{12} u_{n,2} + A_{13} u_{n,3} + A_{14} u_{n,4} - i\mu u_{n,1}|| \le C(\epsilon_n + \delta_n);\\
&||A_{01} u_{n,1} + A_{02} u_{n,2} + A_{03} u_{n,3} + A_{04} u_{n,4} - i\mu u_{n,0}|| \le C(\epsilon_n + \delta_n). && (9.14)
\end{aligned}$$

Since $\dim X_j < \infty$ when $j \ne 0, 3$, subject to a subsequence, we may assume that as $n \to \infty$,

$$u_{n,j} \to u_j, \; j = 1, 2, 4; \quad u_{n,5}, \; u_{n,6} \to 0; \quad u_{n,j} \rightharpoonup u_j, \; j = 0, 3.$$

Passing to the limits in the above inequalities and using the boundedness of A_j and A_{jk} except A_3, we obtain

$$\begin{aligned}
&A_4 u_4 - i\mu u_4 = 0; \quad A_2 u_2 + A_{24} u_4 - i\mu u_2 = 0;\\
&A_1 u_1 + A_{12} u_2 + A_{13} u_3 + A_{14} u_4 - i\mu u_1 = 0;\\
&A_{01} u_1 + A_{02} u_2 + A_{03} u_3 + A_{04} u_4 - i\mu u_0 = 0.
\end{aligned}$$

Moreover, the above inequality involving $A_3 u_{n,3}$ also implies that

$$u_{n,3} \rightharpoonup u_3, \quad A_3 u_{n,3} \rightharpoonup -A_{34} u_4 + i\mu u_3.$$

Since the graph of the closed operator of A_3 as a closed subspace in $X_3 \times X_3$ is also closed under the weak topology, we obtain

$$u_3 \in D(A_3) \text{ and } A_3 u_3 + A_{34} u_4 - i\mu u_3 = 0.$$

These equalities imply that

$$JLu = i\mu u, \quad \text{where } u = u_0 + u_1 + u_2 + u_3 + u_4.$$

In addition, (9.12) implies

$$\langle Lu, u\rangle = 2\langle Lu_1, u_4\rangle + \langle L_2 u_2, u_2\rangle + \langle L_3 u_3, u_3\rangle \leq 0.$$

Due to property (9.9) of $i\mu$, we must have $u = 0$, which immediately yields $u_{n,j} \to 0$, $j = 1, 2, 4, 5, 6$ and thus (9.12) implies $u_{n,3} \to 0$ as well. Then the normalization $||u_n|| = 1$ implies that we must have $\dim \ker L \geq 1$, $||u_{n,0}|| \to 1$ and $u_{n,0} \rightharpoonup 0$. From (9.14), we obtain $\mu = 0$. As $\ker L$ is nontrivial, this again contradicts to (9.9). Therefore Theorem 2.5 holds in this case.

Summarizing the above two cases, **Theorem 2.5 is proved.**

'Structurally unstable' cases. In the following, we will consider cases in Theorem 2.6 for the structural instability. In many applications the symplectic structure J usually does not vary, therefore we will fix J and focus on constructing perturbations to the energy operator L to induce instabilities arising from a purely imaginary eigenvalue $i\mu$ of JL. Recall we have to complexify X, J, and L accordingly. However, keep in mind that we would like to construct real perturbations to create unstable eigenvalues near $i\mu$. This would require the perturbations to also satisfy (A.12), see Remark A.5. Recall that while JL is a linear operator, L and J are complexified as Hermitian forms or anti-linear mappings, see (A.8) and (A.9).

Case 3: $i\mu \in \sigma(JL)$ and $\exists$ a closed subspace $\{0\} \neq Y \subset E_{i\mu}$ such that

$$JL(Y) \subset Y, \text{ and } \langle L\cdot, \cdot\rangle \text{ is non-degenerate and sign indefinite on } Y.$$

REMARK 9.2. *Clearly this includes, but not limited to, the situation where $\langle L\cdot, \cdot\rangle$ is non-degenerate and indefinite on $E_{i\mu}$, a special case of which is when $i\mu$ is isolated in $\sigma(JL)$. It is analyzed in several subcases below.*

We will construct a perturbation $L_\#$ such that $\sigma(JL_\#)$ contains a hyperbolic eigenvalue near $i\mu$. The proof will basically be carried out in some finite dimensional subspaces. Such finite dimensional problems had been well studied in the literature, mostly for the Case 3b below when there are two eigenvectors of opposite signs of $\langle L\cdot, \cdot\rangle$ (see e.g. [**24, 59**]). We could not find a reference for the proof of structural instability when the indefiniteness of $L|_{E_{i\mu}}$ is caused by a Jordan chain of JL (Case 3c below). So we give a detailed proof for the general case, which will also be used in later cases of embedded eigenvalues. Our proof for the Case 3c uses the special basis constructed in Proposition 2.2 for the Jordan blocks of JL on $E_{i\mu}$.

Recall $\bar{u} \in E_{-i\mu}$ for any $u \in E_{i\mu}$. Let

$$Y_\mu = \{u + \bar{v} \mid u, v \in Y\} \subset E_{i\mu} + E_{-i\mu}.$$

From Lemma 3.3 and the assumption on Y, $\langle L\cdot, \cdot\rangle$ is still non-degenerate on Y_μ which is also clearly invariant under JL. Recall that

$$Y_\mu^{\perp_L} = \{u \in X \mid \langle Lu, v\rangle = 0, \ \forall v \in Y_\mu\}.$$

From Lemmas A.2 and 3.2, $Y_\mu^{\perp_L}$ is also invariant under e^{tJL} and $X = Y_\mu \oplus Y_\mu^{\perp_L}$. The definition of Y_μ implies that Y_μ is real, in the sense $\bar{u} \in Y_\mu$ for any $u \in Y_\mu$, and thus the complexification of a real subspace of X. According to Lemma A.3, $JL|_{Y_\mu}$ is a also a Hamiltonian operator satisfying hypotheses (**H1-3**) with the non-degenerate energy L_{Y_μ}, defined in (A.1). Therefore we may apply Proposition 2.2 to Y_μ and $JL|_{Y_\mu}$, where $Y \subset Y_\mu$ is the subspace of all generalized eigenvectors of $i\mu$ of $JL|_{Y_\mu}$. Since L is non-degenerate on Y, it is clear that **Case 3** contains the following three subcases only.

Case 3a: $\mu \neq 0$ *and* $\langle L\cdot, \cdot\rangle$ *changes sign on* $\ker(JL - i\mu) \cap Y$.

In this subcase, let $u_\pm \in \ker(JL - i\mu) \cap Y$ be such that $\pm\langle Lu_\pm, u_\pm\rangle > 0$. By a Gram-Schmidt process, without loss of generality, we may assume

$$\langle Lu_\pm, u_\pm\rangle = \pm 1, \quad \langle Lu_+, u_-\rangle = 0.$$

Note that $u_\pm$ can not be real for $\mu \neq 0$. As we will construct real perturbations to create instability, we have to consider the complex conjugate of $u_\pm$ as well. Let

$$X_1 = span\{u_+, u_-, \overline{u_+}, \overline{u_-}\}, \quad X_2 = X_1^{\perp_L} = \{v \in X \mid \langle Lu_\pm, v\rangle = \langle L\overline{u_\pm}, v\rangle = 0\}.$$

It is clear that subspaces $X_{1,2}$ are comlexifications of real subspaces in the sense

$$\bar{u} \in X_{1,2} \ \text{ if } \ u \in X_{1,2}. \tag{9.15}$$

Note that $\overline{u_\pm}$ are eigenvectors of $-i\mu$ $(\neq i\mu)$ and that $E_{i\mu}$ and $E_{-i\mu}$ are L-orthogonal. Therefore from the complexification process, it is easy to verify that, with respect to this basis of the invariant subspace X_1 of JL, operators L_{X_1} and $JL|_{X_1}$ take the forms

$$L_{X_1} = \begin{pmatrix} \Lambda & 0 \\ 0 & \Lambda \end{pmatrix}, \quad JL|_{X_1} \triangleq A_1 = i\mu \begin{pmatrix} I_{2\times 2} & 0 \\ 0 & -I_{2\times 2} \end{pmatrix},$$

where

$$\Lambda = \begin{pmatrix} 1 & 0 \\ 0 & -1 \end{pmatrix}, \quad I_{2\times 2} = \begin{pmatrix} 1 & 0 \\ 0 & 1 \end{pmatrix}.$$

From the invariance of X_1 and Lemmas A.2 and 3.2, $X_2 = X_1^{\perp_L}$ is also invariant under JL and $X = X_1 \oplus X_2$. In this decomposition, L, JL, and J take the forms

$$L = \begin{pmatrix} L_{X_1} & 0 \\ 0 & L_{X_2} \end{pmatrix}, \quad JL = \begin{pmatrix} A_{X_1} & 0 \\ 0 & A_{X_2} \end{pmatrix}, \quad J = \begin{pmatrix} J_{X_1} & 0 \\ 0 & J_{X_2} \end{pmatrix}, \tag{9.16}$$

where, with respect to the basis of X_1^* dual to $\{u_\pm, \ \overline{u_\pm}\}$,

$$J_{X_1} = i\mu \begin{pmatrix} \Lambda & 0 \\ 0 & -\Lambda \end{pmatrix}, \quad J_{X_2} : X_2^* \supset D(J_{X_2}) \to X_2, \quad J_{X_2}^* = -J_{X_2}.$$

Here, $J^* = -J$ is used.

Consider a perturbation L_1 in the form of

$$L_1 = \begin{pmatrix} L_{1,X_1} & 0 \\ 0 & 0 \end{pmatrix}, \text{ where } L_{1,X_1} = \begin{pmatrix} \epsilon R & 0 \\ 0 & \epsilon R \end{pmatrix}, \quad R = \begin{pmatrix} 0 & 1 \\ 1 & 0 \end{pmatrix}.$$

It is straightforward to verify that L_1 is real, namely, $\langle L_1\bar{u}, \bar{v}\rangle = \overline{\langle L_1 u, v\rangle}$. Let $L_\# = L + L_1$. Clearly the decomposition $X = X_1 \oplus X_2$ is still invariant under $JL_\#$ and orthogonal with respect to $L_\#$. Therefore by a direct computation on the 4×4 matrix $JL_\#|_{X_1}$ which can be further reduced to the 2×2 matrix $i\mu\Lambda(\Lambda + \epsilon R)$, we obtain

$$i\mu \pm \epsilon\mu \in \sigma(JL_\#).$$

Therefore $\sigma(JL_{\#})$ contains hyperbolic eigenvalues near $i\mu$ for any $\epsilon \neq 0$.
Case 3b: $\mu = 0$ and $\langle L\cdot, \cdot\rangle$ changes sign on $\ker(JL - i\mu) \cap Y$.

In this case one may proceed as in the above through (9.16), with $J_{X_1} = 0$ however. Therefore no hyperbolic eigenvalue can bifurcate through such type of perturbations of L.

Cases 3c: $\mu \neq 0$ and Y contains a non-trivial Jordan chain $u_j = (JL - i\mu)^{j-1}u_1$, $j = 1, \dots, k > 1$, of JL such that $u_k \in \ker(JL - i\mu)\backslash\{0\}$ and $\langle L\cdot, \cdot\rangle$ is non-degenerate on span$\{u_1, \dots, u_k\}$.

Again in this case let

$$X_1 = span\{u_j,\ \overline{u_j} \mid j = 1, \dots, k\}, \quad X_2 = X_1^{\perp_L},$$

which is a L-orthogonal invariant decomposition under JL satisfying (9.15) and thus the forms (9.16) hold. From Proposition 2.2, without loss of generality, we may assume that, with respect to the basis $\{u_1, \dots, u_k, \overline{u_1}, \dots, \overline{u_k}\}$ (as well as its dual basis in X_1^*), L_{X_1}, $JL|_{X_1} \triangleq A_{X_1}$, and J_{X_1} take the forms

$$L_{X_1} = \begin{pmatrix} B_+ & 0 \\ 0 & \overline{B_+} \end{pmatrix}, \quad A_{X_1} = \begin{pmatrix} A_+ & 0 \\ 0 & \overline{A_+} \end{pmatrix}, \quad J_{X_1} = \begin{pmatrix} J_+ & 0 \\ 0 & \overline{J_+} \end{pmatrix},$$

where

$$B_+ = \begin{pmatrix} 0 & \dots & 0 & b_k \\ 0 & \dots & b_{k-1} & 0 \\ \dots & & & \\ b_1 & \dots & 0 & 0 \end{pmatrix}, \quad A_+ = \begin{pmatrix} i\mu & 0 & \dots & 0 & 0 \\ 1 & i\mu & \dots & 0 & 0 \\ \dots & & & & \\ 0 & 0 & \dots & 1 & i\mu \end{pmatrix},$$

and $b_{j+1} = -b_j$, $b_{k+1-j} = \overline{b_j}$. Therefore $b_j \in \{\pm i\}$ if $2|k$ or $b_j \in \{\pm 1\}$ otherwise. Then one may compute

$$J_+ = \begin{pmatrix} 0 & 0 & \dots & 0 & i\mu\overline{b_1}^{-1} \\ 0 & 0 & \dots & i\mu\overline{b_2}^{-1} & \overline{b_1}^{-1} \\ \dots & & & & \\ i\mu\overline{b_k}^{-1} & \overline{b_{k-1}}^{-1} & \dots & 0 & 0 \end{pmatrix}.$$

Here note that $\overline{b_j}^{-1}$ instead of b_j^{-1} appears in above J_+, namely

$$J_+(u_j^*) = i\mu\overline{b_{k+1-j}}^{-1}u_{k+1-j} + \overline{b_{k+1-j}}^{-1}u_{k+2-j}$$

where $u_{k+1} = 0$ is understood. This is due to the *anti-linear* complexification of L and J, see (A.9). In fact, let $\{u_j^*,\ \overline{u_j}^* \mid j = 1, \dots, k\}$ be the dual basis in X_1^* which are complex linear functionals. We have

$$\begin{aligned} \langle u_l^*, Ju_j^*\rangle =& \langle u_l^*, J(b_{k+1-j}^{-1}Lu_{k+1-j})\rangle = \langle u_l^*, \overline{b_{k+1-j}}^{-1}JLu_{k+1-j})\rangle \\ =& \overline{b_{k+1-j}}^{-1}\langle u_l^*, JLu_{k+1-j}\rangle = \overline{b_{k+1-j}}^{-1}\langle u_l^*, i\mu u_{k+1-j} + u_{k+2-j}\rangle \\ =& \overline{b_{k+1-j}}^{-1}\left(i\mu\delta_{l,k+1-j} + \delta_{l,k+2-j}\right). \end{aligned}$$

Consider perturbations in the form of

$$L_1 = \begin{pmatrix} L_{1,X_1} & 0 \\ 0 & 0 \end{pmatrix}, \quad \text{where } L_{1,X_1} = \epsilon\begin{pmatrix} B & 0 \\ 0 & \overline{B} \end{pmatrix}, \quad B = \begin{pmatrix} 0 & \dots & 0 & 0 \\ \dots & & & \\ 0 & \dots & 0 & 0 \\ 0 & \dots & 0 & 1 \end{pmatrix}.$$

Clearly L_1 is real in the sense $\langle L_1\bar{u},\bar{v}\rangle = \overline{\langle L_1 u, v\rangle}$. Let $L_\# = L + L_1$ and the decomposition $X = X_1 \oplus X_2$ is still invariant under $JL_\#$ and orthogonal with respect to $L_\#$. Therefore $\sigma\big(J_+(B_+ + \epsilon B)\big) \subset \sigma(JL_\#)$. By direct computation, we obtain the matrix

$$J_+(B_+ + \epsilon B) = \begin{pmatrix} i\mu & 0 & \dots & 0 & i\epsilon\mu\overline{b_1}^{-1} \\ 1 & i\mu & \dots & 0 & \epsilon\overline{b_1}^{-1} \\ \dots & & & & \\ 0 & 0 & \dots & 1 & i\mu \end{pmatrix}$$

and its characteristic polynomial

$$\det\big(\lambda - J_+(B_+ + \epsilon B)\big) = (-i)^k p\big(i(\lambda - i\mu)\big),$$

where

$$p(\lambda) = \lambda^k - \epsilon b\lambda + \epsilon b\mu, \quad b = (-i)^{k-1}\overline{b_1}^{-1} \in \{\pm 1\}.$$

To find hyperbolic eigenvalues of $J_+(B_+ + \epsilon B)$, it is equivalent to show that $p(\lambda) = 0$ has a root $\lambda \notin \mathbf{R}$. Choose the sign of ϵ such that $\epsilon b\mu > 0$. Denote $c_1, \cdots, c_k$ to be all the $k-$th roots of $-|b\mu|$, which are not real except for at most one. So we can assume $\operatorname{Im} c_1 \neq 0$. Let $\delta = |\epsilon|^{\frac{1}{k}}$. Then $p(\lambda) = 0$ is equivalent to

$$\left(\frac{\lambda}{\delta}\right)^k - \delta|b|\frac{|\mu|}{\mu}\left(\frac{\lambda}{\delta}\right) + |b\mu| = 0. \tag{9.17}$$

When $\delta \ll 1$, by the Implicit Function Theorem, (9.17) has k roots of the form

$$\frac{\lambda_j}{\delta} = c_j + O(\delta), \ j = 1, \cdots, k,$$

among which $\lambda_1 = \delta c_1 + O(\delta^2)$ satisfies $\operatorname{Im}\lambda_1 \neq 0$. This implies that $J_+(B_+ + \epsilon B)$ has a hyperbolic eigenvalue of the form $i\mu - i\delta c_1 + O(\delta^2)$.

Cases 3d: $\mu = 0$ and Y contains a non-trivial Jordan chain of length ≥ 3. Let $u_j = (JL)^{j-1}u_1$, $j = 1, \dots, k, (k \geq 3)$ be a Jordan chain of JL such that $\langle L\cdot,\cdot\rangle$ is non-degenerate on $span\{u_1, \dots, u_k\}$.

In this case one may proceed as in the above with

$$p(\lambda) = \lambda^k - \epsilon b\lambda = \lambda(\lambda^{k-1} - \epsilon b).$$

Choose ϵ such that $\epsilon b < 0$. Since $k \geq 3$, $p(\lambda)$ has a complex root which implies that $J_+(B_+ + \epsilon B)$ has a hyperbolic eigenvalue. For $\mu = 0$ and $k = 2$, by straightforward computations, it can be shown that J must be degenerate and $J_+(B_+ + \epsilon B)$ has only eigenvalue 0 and a purely imaginary eigenvalues for any 2×2 Hermitian matrix B and $\epsilon \ll 1$.

Case 4: $i\mu \in \sigma(JL) \cap i\mathbf{R}\backslash\{0\}$ and $\langle L\cdot,\cdot\rangle$ is degenerate on $E_{i\mu} \neq \{0\}$.

In this case, Proposition 2.3 implies that $i\mu$ must be non-isolated in $\sigma(JL)$ and we start with the following lemma to isolate $i\mu$ through a perturbation.

LEMMA 9.5. *Assume **(H1-3)**. Suppose $i\mu \in \sigma(JL) \cap i\mathbf{R}$ is non-isolated in $\sigma(JL)$. For any $\epsilon > 0$, there exists a symmetric bounded linear operator $L_1 : X \to X^*$ satisfying* (A.12) *such that $|L_1| < \epsilon$ and $i\mu \in \sigma(JL_\#)$ is an isolated eigenvalue, where $L_\# = L + L_1$, and $\langle L_\# u, u\rangle > 0$ for some generalized eigenvector u of the eigenvalue $i\mu$ of $JL_\#$.*

PROOF. Since $\sigma(JL)$ is symmetric about both real and imaginary axes, without loss of generality we can assume that $\mu \geq 0$.

Let $X = \Sigma_{j=0}^{6} X_j$ be the decomposition given in Theorem 2.1 with associated projections P_j. We will use the notations there in the rest of the proof. Recall Theorem 2.1 is proved without the complexification, i.e. in the framework of real Hilbert space X and real operators J and L, the resulted decomposition and operators are real. After the complexification, X_j are real in the sense of (9.15) and the operators satisfy (A.12) and the blocks in L and J are anti-linear.

As $i\mu \in \sigma(JL)$ is assumed to be non-isolated and $\dim X_j < \infty$, $j \neq 0, 3$, it must hold $i\mu \in \sigma(A_3)$. Since A_3 is anti-self-adjoint with respect to the positive definite Hermitian form $\langle L_{X_3}\cdot, \cdot\rangle$, it induces a resolution of the identity. Namely there exists a family of projections $\{\Pi_\lambda\}_{\lambda \in \mathbf{R}}$ on X_3 such that

(1) $\lim_{\lambda \to \lambda_0 +} \Pi_\lambda u = \Pi_{\lambda_0} u$, for all $\lambda_0 \in \mathbf{R}$ and $u \in X_3$;
(2) $\Pi_{\lambda_1}\Pi_{\lambda_2} = \Pi_{\min\{\lambda_1, \lambda_2\}}$, for all $\lambda_{1,2} \in \mathbf{R}$;
(3) $\langle L_{X_3}\Pi_\lambda u_1, u_2\rangle = \langle L_{X_3} u_1, \Pi_\lambda u_2\rangle$ for any $u_{1,2} \in X_3$ and $\lambda \in \mathbf{R}$;
(4) $u = \int_{-\infty}^{+\infty} d\Pi_\lambda u$, $A_3 u = \int_{-\infty}^{+\infty} i\lambda \, d\Pi_\lambda u$, for any $u \in X_3$;
(5) $d\Pi_\lambda = \overline{d\Pi_{-\lambda}}$ for any $\lambda \in \mathbf{R}$.

Here the last property is due to the fact that J and L are real satisfying (A.12).

For $\mu > 0$, define a perturbation $\tilde{L}_1 : X_3 \to X_3^*$ by

$$\tilde{L}_1 u = L_{X_3}\Big(\int_{|\lambda-\mu|<\nu} \frac{\lambda-\mu}{\lambda} d\Pi_\lambda u + \int_{|\lambda+\mu|<\nu} \frac{\lambda+\mu}{\lambda} d\Pi_\lambda u\Big), \quad \forall u \in X_3,$$

where $\nu \in (0, \mu)$ is a small constant to be determined later. As $\Pi_\lambda u$ is continuous from the right, the integrals take the same values on the open intervals or half open half closed interval like $[\mu - \nu, \mu + \nu)$.

For $\mu = 0$, define

$$\tilde{L}_1 u = L_{X_3} \int_{-\nu}^{\nu} d\Pi_\lambda u, \quad \forall u \in X_3$$

where again $\nu > 0$ is determined later.

For $\mu > 0$, like L_{X_3}, it is clear that $\tilde{L}_1$ is anti-linear satisfying (A.9). We will verify that $\tilde{L}_1$ is also real and symmetric. For any $u \in X$, one may compute using $d\Pi_\lambda = \overline{d\Pi_{-\lambda}}$

$$\begin{aligned}
\overline{\tilde{L}_1 u} =& \overline{L_{X_3}\Big(\int_{|\lambda-\mu|<\nu} \frac{\lambda-\mu}{\lambda} d\Pi_\lambda u + \int_{|\lambda+\mu|<\nu} \frac{\lambda+\mu}{\lambda} d\Pi_\lambda u\Big)} \\
=& L_{X_3}\Big(\int_{|\lambda-\mu|<\nu} \frac{\lambda-\mu}{\lambda} \overline{d\Pi_\lambda}\bar{u} + \int_{|\lambda+\mu|<\nu} \frac{\lambda+\mu}{\lambda} \overline{d\Pi_\lambda}\bar{u}\Big) \\
=& L_{X_3}\Big(\int_{|\lambda-\mu|<\nu} \frac{\lambda-\mu}{\lambda} d\Pi_{-\lambda}\bar{u} + \int_{|\lambda+\mu|<\nu} \frac{\lambda+\mu}{\lambda} d\Pi_{-\lambda}\bar{u}\Big).
\end{aligned}$$

Through a change of variable $\lambda \to -\lambda$, we obtain $\overline{\tilde{L}_1 u} = \tilde{L}_1 \bar{u}$, namely, $\tilde{L}_1$ is real (the complxification of a real linear operator). Moreover, for any $u_{1,2} \in X$, we have

$$\begin{aligned}
\langle \tilde{L}_1 u_1, u_2\rangle &= \langle L_{X_3}\Big(\int_{|\lambda-\mu|<\nu} \frac{\lambda-\mu}{\lambda} d\Pi_\lambda u_1 + \int_{|\lambda+\mu|<\nu} \frac{\lambda+\mu}{\lambda} d\Pi_\lambda u_1\Big), u_2\rangle \\
&= \int_{|\lambda-\mu|<\nu} \frac{\lambda-\mu}{\lambda} d\langle L_{X_3}\Pi_\lambda u_1, u_2\rangle + \int_{|\lambda+\mu|<\nu} \frac{\lambda+\mu}{\lambda} d\langle L_{X_3}\Pi_\lambda u_1, u_2\rangle.
\end{aligned}$$

Since L_{X_3} is Hermitian and $\langle \Pi_\lambda \cdot, \cdot \rangle = \langle \cdot, \Pi_\lambda \cdot \rangle$ on X_3, we obtain that $\tilde{L}_1$ is Hermitian. Therefore $\left\langle \tilde{L}_1 \cdot, \cdot \right\rangle$ is the complexification of a real bounded symmetric quadratic form on X_3. Clearly, in the equivalent norm $\langle L_{X_3} u, u \rangle^{\frac{1}{2}}$ on X_3,

$$|\tilde{L}_1| \leq \frac{\nu}{\mu - \nu} \to 0, \qquad \text{as } \nu \to 0.$$

The same properties also hold for $\tilde{L}_1$ for $\mu = 0$ and we skip the details.

Let

$$L_1 = P_3^* \tilde{L}_1 P_3, \quad L_\# = L - L_1.$$

Accordingly in this decomposition

$$L_\# \longleftrightarrow \begin{pmatrix} 0 & 0 & 0 & 0 & 0 & 0 & 0 \\ 0 & 0 & 0 & 0 & B_{14} & 0 & 0 \\ 0 & 0 & L_{X_2} & 0 & 0 & 0 & 0 \\ 0 & 0 & 0 & L_{X_3} + \tilde{L}_1 & 0 & 0 & 0 \\ 0 & B_{14}^* & 0 & 0 & 0 & 0 & 0 \\ 0 & 0 & 0 & 0 & 0 & 0 & B_{56} \\ 0 & 0 & 0 & 0 & 0 & B_{56}^* & 0 \end{pmatrix}.$$

From Corollary 2.1, one can compute

$$JL_\# \longleftrightarrow \begin{pmatrix} 0 & A_{01} & A_{02} & A_{03}(I - L_{X_3}^{-1}\tilde{L}_1) & A_{04} & 0 & 0 \\ 0 & A_1 & A_{12} & A_{13}(I - L_{X_3}^{-1}\tilde{L}_1) & A_{14} & 0 & 0 \\ 0 & 0 & A_2 & 0 & A_{24} & 0 & 0 \\ 0 & 0 & 0 & A_3(I - L_{X_3}^{-1}\tilde{L}_1) & A_{34} & 0 & 0 \\ 0 & 0 & 0 & 0 & A_4 & 0 & 0 \\ 0 & 0 & 0 & 0 & 0 & A_5 & 0 \\ 0 & 0 & 0 & 0 & 0 & 0 & A_6 \end{pmatrix}.$$

Due to the upper triangular structure of $JL_\#$ and the finite dimensionality of $X_{1,2}$, in order to prove that $i\mu$ belongs to and is isolated in $\sigma(JL_\#)$, it suffices to show that $i\mu$ belongs to and is isolated in $\sigma\big(A_3(I - L_{X_3}^{-1}\tilde{L}_1)\big)$. In fact, for any $u \in X_3$,

$$A_3(I - L_{X_3}^{-1}\tilde{L}_1)u = i \int_S \mu \, d\Pi_\lambda u + i \int_{\mathbf{R}\setminus S} \lambda \, d\Pi_\lambda u, \tag{9.18}$$

where $S = (-\mu - \nu, -\mu + \nu) \cup (\mu - \nu, \mu + \nu)$. Since Π_λ is not constant on S as $i\mu \in \sigma(A_3)$, we obtain that $i\mu$ is an isolated eigenvalue of $A_3(I - L_{X_3}^{-1}\tilde{L}_1)$ and thus of $\sigma(JL_\#)$ as well. Indeed, for any $u \in R\,(\Pi_{\mu+\nu} - \Pi_{\mu-\nu})$, by (9.18) we have

$$A_3(I - L_{X_3}^{-1}\tilde{L}_1)u = i\mu u.$$

So $i\mu$ is an eigenvalues of $A_3(I - L_{X_3}^{-1}\tilde{L}_1)$. To show $i\mu$ is isolated, taking any $\alpha \in \mathbf{C}$ such that $0 < |\alpha - i\mu| < \nu$, then we have

$$\left(\alpha - A_3(I - L_{X_3}^{-1}\tilde{L}_1)\right)^{-1} = \int_S (\alpha - i\mu)^{-1} \, d\Pi_\lambda + \int_{\mathbf{R}\setminus S} (\alpha - i\lambda)^{-1} \, d\Pi_\lambda,$$

which is clearly a bounded operator.

Finally, we prove that there exists a generalized eigenvector u of $i\mu$ of $JL_\#$ such that $\langle L_\# u, u \rangle > 0$. Since $\dim X_1 < \infty$, there exists an integer $K > 0$ such that

$$X_1 = Y_\mu \oplus \tilde{Y}, \text{ where } Y_\mu = \ker(A_1 - i\mu)^K \cap X_1, \; \tilde{Y} = (A_1 - i\mu)^K X_1.$$

In the following we proceed in the case of $\mu > 0$ first. Let

$$Z_\mu = \{u - (-i\mu)^{-K} P_0 (JL - i\mu)^K u \mid u \in Y_\mu\}, \quad \tilde{Z} = \ker L \oplus \tilde{Y}.$$

Note that the upper triangular structure of JL implies that

$$(A_1 - i\mu)^K = P_1 (JL - i\mu)^K|_{X_1}.$$

Using this observation and the invariance of $\tilde{X} = \ker L \oplus X_1$ under JL, we obtain through straightforward computations

$$\tilde{X} = Z_\mu \oplus \tilde{Z}, \; Z_\mu = \ker(JL - i\mu)^K \cap \tilde{X}, \; \tilde{Z} = (JL - i\mu)^K \tilde{X}, \tag{9.19}$$

and on the invariant subspaces Z_μ and $\tilde{Z}$

$$\sigma(JL|_{Z_\mu}) = \{i\mu\} \text{ if } i\mu \in \sigma(A_1), \quad i\mu \notin \sigma(JL|_{\tilde{Z}}). \tag{9.20}$$

Let $P : \tilde{X} \to \tilde{Z}$ be the projection associated to the above decomposition and $u_3 \in X_3$ be such that

$$A_3(I - L_{X_3}^{-1} \tilde{L}_1) u_3 = i\mu u_3.$$

The structure of $JL_\#$ implies $(JL_\# - i\mu) u_3 \in \tilde{X}$. Let

$$\tilde{u} = \big((JL - i\mu)|_{\tilde{Z}}\big)^{-1} \tilde{P}(JL_\# - i\mu) u_3 \in \tilde{Z} \subset \tilde{X}, \quad u = u_3 - \tilde{u}.$$

By using $(L_\# - L)|_{\tilde{X}} = 0$, it is easy to verify that

$$(JL_\# - i\mu) u \in Z_\mu,$$

which implies

$$(JL_\# - i\mu)^{K+1} u = 0.$$

From the structure of $L_\#$, straightforward computation leads to

$$\langle L_\# u, u\rangle = \langle (L_3 + \tilde{L}_1) u_3, u_3\rangle > 0,$$

for $0 < \nu << 1$.

The case of $\mu = 0$ is largely similar. Let

$$Z_0 = Y_0 \oplus \ker L, \quad \tilde{Z} = \tilde{Y}$$

and (9.19) and (9.20) still hold. The rest of the argument follows in exactly the same procedure. □

We return to construct a perturbation L_1 to L to create unstable eigenvalues. In Case 4, E^D in Proposition 2.2 is non-trivial and finite dimensional, therefore

$$\exists\, 0 \neq u_0 \in \ker(JL - i\mu) \text{ such that } \langle L u_0, u_0\rangle = 0, \tag{9.21}$$

where $u_0 \in E^D$. Since $\mu \neq 0$ implies $\overline{u_0} \in E_{-i\mu}$ with $\langle L u_0, \overline{u_0}\rangle = 0$, let

$$Y_0 = span\{u_0, \overline{u_0}\} \subset \ker(JL - i\mu) \oplus \ker(JL + i\mu). \tag{9.22}$$

The following decomposition lemma is our first step in the construction of a hyperbolically generating perturbation.

LEMMA 9.6. *Suppose $0 \neq i\mu \in \sigma(JL) \cap i\mathbf{R}$ satisfying (9.21). Let Y_0 be defined in (9.22). Then there exists $w \in D(JL)$ with $\overline{w} \neq w$ and a codim-4 closed subspace $Y_1 \subset X$ satisfying (9.15) such that $X = Y_0 \oplus Y_1 \oplus Y_2$, where $Y_2 = span\{w, \overline{w}\}$.*

Moreover, in this decomposition and the bases $\{u_0, \overline{u_0}\}$, $\{w, \bar{w}\}$ *on* $Y_{0,2}$ *respectively,* L *and* JL *take the forms*

$$L \longleftrightarrow \begin{pmatrix} 0 & 0 & I_{2\times 2} \\ 0 & L_{Y_1} & 0 \\ I_{2\times 2} & 0 & 0 \end{pmatrix}, \quad JL \longleftrightarrow \begin{pmatrix} i\mu\Lambda & A_{01} & A_{02} \\ 0 & A_1 & A_{12} \\ 0 & 0 & i\mu\Lambda \end{pmatrix}, \quad \Lambda = \begin{pmatrix} 1 & 0 \\ 0 & -1 \end{pmatrix}.$$

Here all blocks are bounded operators except $A_1 = J_{Y_1} L_{Y_1}$ *and* (Y_1, J_{Y_1}, L_{Y_1}) *satisfies* ***(H1-3)***.

Proof. Let

$$\tilde{Y} = \{u \in X \mid \langle Lu, u_0 \rangle = 0 = \langle Lu, \overline{u_0} \rangle\} \supset \{u_0, \overline{u_0}\}.$$

Clearly $\tilde{Y}$, satisfying (9.15), is the complexification of some real codim-2 subspace. Lemma 3.2 implies that $\tilde{Y}$ is invariant under JL. Let

$$\tilde{Y}_1 = \{u \in \tilde{Y} \mid (u, u_0) = (u, \overline{u_0}) = 0\}.$$

Since $Y_0 \cap \ker L = \{0\}$ and $D(JL)$ is dense in X, there exists a 2-dim subspace $\tilde{Y}_2 \subset D(JL)$ such that $\langle Lu, v \rangle$, $u \in Y_0$ and $v \in \tilde{Y}_2$, defines a non-degenerate bilinear form on $Y_0 \otimes \tilde{Y}_2$. Clearly we have $X = Y_0 \oplus \tilde{Y}_1 \oplus \tilde{Y}_2$ and in this decomposition L takes the form

$$L \longleftrightarrow \begin{pmatrix} 0 & 0 & B_{02} \\ 0 & L_{Y_1} & B_{12} \\ B_{02}^* & B_{12}^* & B_{22} \end{pmatrix},$$

where $B_{02} : \tilde{Y}_2 \to Y_0^*$ is non-degenerate and $B_{22}^* = B_{22}$. Through exactly the same procedure as in the proof of Proposition 6.1, we may obtain subspaces Y_1 and $\tilde{X}_2$ as graphs of bounded linear operators from $\tilde{Y}_{1,2}$ to Y_0 such that $X = Y_0 \oplus Y_1 \oplus X_2$ and in this decomposition L takes the form

$$L \longleftrightarrow \begin{pmatrix} 0 & 0 & B \\ 0 & L_{Y_1} & 0 \\ B^* & 0 & 0 \end{pmatrix},$$

where $B : X_2 \to Y_0^*$ is non-degenerate. There exists $w \in X_2 \subset D(JL)$ such that $\langle Lu_0, w \rangle = 1$ and $\langle Lu_0, \bar{w} \rangle = \langle L\overline{u_0}, w \rangle = 0$, which also implies $\langle L\overline{u_0}, \bar{w} \rangle = 0$, where (A.12) is used. Let $Y_2 = span\{w, \bar{w}\}$. From the definition of w, $\tilde{Y}$, and Y_1, we have $X = Y_0 \oplus Y_1 \oplus Y_2$, associated with projections $P_{0,1,2}$, and in this decomposition, the desired block form of L is achieved. Applying Lemma A.3 to $X = (Y_0 \oplus Y_2) \oplus Y_1$, we obtain that (Y_1, J_{Y_1}, L_{Y_1}) satisfies (**H1-3**), where $J_{Y_1} = P_1 J P_1^*$. The upper triangular block form of JL is due to the invariance of Y_0 and $\tilde{Y} = Y_0 \oplus Y_2$.

To complete the proof of the lemma, we are left to show $P_2 JLw = i\mu w$, which along with the facts that $\tilde{Y}$ satisfies (9.15) and JL satisfies (A.12) also implies $P_2 JL\bar{w} = -i\mu\bar{w}$. From $(JL)^* = -LJ$ (Corollary A.1), we have

$$\langle LJLw, u_0 \rangle = -\langle Lw, JLu_0 \rangle = i\mu \langle Lw, u_0 \rangle = i\mu$$

and similarly $\langle LJLw, \overline{u_0} \rangle = -i\mu \langle Lw, \overline{u_0} \rangle = 0$. According to the definitions of $\tilde{Y}$ and w, we obtain $P_2 JLw = i\mu w$ and the lemma is proved. □

With the above lemmas, we are ready to construct a perturbed energy operator $L_\#$ to create unstable eigenvalues of $JL_\#$ near $i\mu$ in the Case 4. We start with the decomposition given in Lemma 9.6. Since $i\mu$ is an eigenvalue of JL non-isolated in $\sigma(JL)$, we have $i\mu \in \sigma(J_{Y_1} L_{Y_1})$ and is non-isolated in $\sigma(J_{Y_1} L_{Y_1})$. From Lemma 9.5, there exists a sufficiently small symmetric bounded linear operator $\tilde{L}_2 : Y_1 \to Y_1^*$ such that $i\mu \in \sigma\big(J_{Y_1}(L_{Y_1} + \tilde{L}_2)\big)$ and is isolated with an eigenvector $u_1 \in Y_1$ satisfying $\langle (L_{Y_1} + \tilde{L}_2)u_1, u_1\rangle > 0$. Let $L_2 = P_1^*\tilde{L}_2 P_1$ and $\tilde{L}_\# = L + L_2$, then the block forms of $L_\#$ and $J\tilde{L}_\#$ imply that $i\mu \in \sigma(J\tilde{L}_\#)$ is isolated and

$$u_0, u_1 \in \ker(J\tilde{L}_\# - i\mu), \quad \langle \tilde{L}_\# u_0, u_0\rangle = 0, \quad \text{and } \langle \tilde{L}_\# u_1, u_1\rangle > 0. \tag{9.23}$$

Since $i\mu$ is isolated in $\sigma(J\tilde{L}_\#)$, Proposition 2.3 implies that $\langle \tilde{L}_\# \cdot, \cdot\rangle$ is non-degenerate on $E_{i\mu}(J\tilde{L}_\#)$, the subspace of generalized eigenvectors of $i\mu$ for $J\tilde{L}_\#$. Moreover, by (9.23), $\langle \tilde{L}_\# \cdot, \cdot\rangle$ is sign indefinite on $E_{i\mu}(J\tilde{L}_\#)$. This situation has been covered in **Case 3**. Therefore there exists a sufficient small symmetric bounded linear operator $L_3 : X \to X^*$ such that there exists $\lambda \in \sigma(JL_\#)\backslash i\mathbf{R}$ sufficiently close to $i\mu$, where $L_\# = L + L_2 + L_3$.

Case 5: $i\mu \in \sigma(JL) \cap i\mathbf{R}\backslash\{0\}$ is non-isolated and $\langle L\cdot, \cdot\rangle$ is negative definite on $E_{i\mu} \neq \{0\}$.

Much as in **Case 4** (but more easily), we can construct sufficiently small symmetric bounded perturbations to the energy operator L to create unstable eigenvalues. In fact, Proposition 2.2 implies that in Case 5, it holds $\ker(JL - i\mu) = E_{i\mu}$. Let

$$Y_0 = E_{i\mu} \oplus E_{-i\mu}, \quad Y = Y_0^{\perp_L} = \{v \in X \mid \langle Lv, u\rangle = \langle Lv, \overline{u}\rangle = 0, \ \forall u \in E_{i\mu}\}.$$

Since $\langle L\cdot, \cdot\rangle$ is negative on Y_0, Lemma A.2 implies that $X = Y_0 \oplus Y$ associated with projections $P_{Y_0, Y}$. In this decomposition L and JL take the forms

$$L \longleftrightarrow \begin{pmatrix} L_{Y_0} & 0 \\ 0 & L_Y \end{pmatrix}, \quad JL \longleftrightarrow \begin{pmatrix} A_0 & 0 \\ 0 & A \end{pmatrix},$$

where A_0 is a bounded operator satisfying $A^2 + \mu^2 = 0$. Lemma A.3 implies that $A = J_Y L_Y$ and (Y, J_Y, L_Y) satisfies **(H1-3)**. Clearly it still holds that $i\mu \in \sigma(J_Y L_Y)$ and is non-isolated there. Applying Lemma 9.5 to L_Y, we obtain a perturbation $\tilde{L} : Y \to Y^*$ such that $i\mu$ is an isolated point in $\sigma\big(J_Y(L_Y + \tilde{L})\big)$. Let $\tilde{L}_\# = L + P_Y^*\tilde{L}P_Y$ and we obtain that $i\mu$ is an isolated point in $\sigma(J\tilde{L}_\#)$ with $\langle \tilde{L}_\# \cdot, \cdot\rangle$ sign indefinite on its eigenspace. This is a case covered in **Case 3** and thus there exists a sufficient small symmetric bounded linear perturbation $L_\#$ to L so that $JL_\#$ has an unstable eigenvalue close to $i\mu$.

Proof of Theorem 2.6. It suffices to show that **Cases 3, 4, 5** cover all the cases in Theorem 2.6. In fact, if $\langle L\cdot, \cdot\rangle$ is degenerate on $E_{i\mu} \neq \{0\}$ and $\mu \neq 0$, this is precisely **Case 4**. Let us consider the case when $\langle L\cdot, \cdot\rangle$ is non-degenerate on $E_{i\mu}$ and satisfies the assumptions in Theorem 2.6. Then $\langle L\cdot, \cdot\rangle$ is either sign indefinite on $E_{i\mu}$ (**Case 3**) or is negative definite on $E_{i\mu}$ for an eigenvalue $i\mu \neq 0$ non-isolated in $\sigma(JL)$ (**Case 5**). $\square$

CHAPTER 10

Proof of Theorem 2.7 where (H2.b) is weakened

In this chapter, we consider the case when (**H2.b**) is weakened, namely, L is only assumed to be positive on X_+, but not necessarily uniformly positive. More precisely, we will prove Theorem 2.7 under hypotheses (**B1-5**) given in Section 2.6. In Section 11.6, as an example we will consider the stability of traveling waves of a nonlinear Schrödinger equation with non-vanishing condition at infinity in two dimensions.

Initial decomposition of the phase space. We adopt the notations as in Chapter 3. Let $P_{\pm,0} : X \to X_{\pm,0}$ be the projections associated to the decomposition $X = X_- \oplus \ker L \oplus X_+$, where $X_0 = \ker L$, and

$$\tilde{X}^*_{\pm,0} = P^*_{\pm,0} X^*_{\pm,0} \subset X^*.$$

We also let

$$X_{\leq 0} = X_- \oplus \ker L, \quad P_{\leq 0} = P_0 + P_- = I - P_+, \quad \tilde{X}^*_{\leq 0} = \tilde{X}^*_- \oplus \tilde{X}^*_0.$$

Clearly we have

$$\tilde{X}^*_+ = \ker i^*_{X_{\leq 0}}, \quad \tilde{X}^*_{\leq 0} = \ker i^*_{X_+} \subset Q_0(X), \quad X^* = \tilde{X}^*_{\leq 0} \oplus \tilde{X}^*_+, \tag{10.1}$$

where assumption (**B5**) is used. Since $\langle Lu, u \rangle < 0$ on $X_- \backslash \{0\}$ and $\dim X_- = n^-(L) < \infty$, there exists $\delta > 0$ such that

$$\langle Lu, u \rangle \leq -\delta \|u\|^2, \quad \forall\, u \in X_-.$$

From (**B4**), we also have

$$LX_+ \subset \tilde{X}^*_+, \quad LX_{\leq 0} = \tilde{X}^*_- \subset \tilde{X}^*_{\leq 0}.$$

Denote

$$L_+ = i^*_{X_+} L i_{X_+} : X_+ \to X^*_+, \quad L_{\leq 0} = i^*_{X_{\leq 0}} L i_{X_{\leq 0}} : X_{\leq 0} \to X^*_{\leq 0},$$

which along with the L-orthogonality in (**B4**) implies

$$L = P^*_+ L_+ P_+ + P^*_{\leq 0} L_{\leq 0} P_{\leq 0}.$$

While the decomposition is not necessarily Q_0-orthogonal, we have the following lemma. Let

$$Q_0^{\leq 0,+} = i^*_{\leq 0} Q_0 i_{X_+} : X_+ \to X^*_{\leq 0}, \quad Q_0^{+,\leq 0} = i^*_{X_+} Q_0 i_{\leq 0} : X_{\leq 0} \to X^*_+,$$
$$Q_0^{\leq 0} = i^*_{X_{\leq 0}} Q_0 i_{X_{\leq 0}} : X_{\leq 0} \to X^*_{\leq 0}, \quad Q_0^+ = i^*_{X_+} Q_0 i_{X_+} : X_+ \to X^*_+.$$

Clearly $Q_0^{\leq 0,+} = (Q_0^{+,\leq 0})^*$ and in the decomposition $X = X_{\leq 0} \oplus X_+$ and $X^* = P^*_{\leq 0} X^*_{\leq 0} \oplus P^*_+ X^*_+$, operator Q_0 takes the form $\begin{pmatrix} Q_0^{\leq 0} & Q_0^{\leq 0,+} \\ Q_0^{+,\leq 0} & Q_0^+ \end{pmatrix}$. Since $\langle Q_0 u, u \rangle > 0$ for all $0 \neq u \in X$, Q_0^+ and $Q_0^{\leq 0}$, as well as L_+, are bounded, symmetric, and positive. Therefore $Q_0^{\leq 0} : X_{\leq 0} \to X^*_{\leq 0}$ and $Q_0^+, L_+ : X_+ \to X^*_+$ are injective with dense ranges. Consequently, $(Q_0^{\leq 0})^{-1} : X_{\leq 0}^* \to X_{\leq 0}$ and $(Q_0^+)^{-1}, L_+^{-1} : X_+^* \to X_+$ are densely defined, closed, and positive operators with

$$\left((Q_0^{\leq 0})^{-1}\right)^* = (Q_0^{\leq 0})^{-1}, \left((Q_0^+)^{-1}\right)^* = (Q_0^+)^{-1},$$

and $(L_+^{-1})^* = L_+^{-1}$.

LEMMA 10.1. *It holds that $P_+^* Q_0^+(X_+) \subset \tilde{X}_+^*$ is dense in $\tilde{X}_+^*$ and*

$$Q_0(X) = \tilde{X}_{\leq 0}^* \oplus P_+^* Q_0^+(X_+), \quad P_+^* Q_0^+(X_+) = Q_0(X) \cap \tilde{X}_+^*,$$

with $(Q_0^{\leq 0})^{-1}$ and $(Q_0^+)^{-1} Q_0^{+,\leq 0}$ being bounded operators. Moreover,

$$A \triangleq Q_0^{-1} P_+^* Q_0^+ : X_+ \to X_2, \quad \textit{where } X_2 = Q_0^{-1}(\tilde{X}_+^*) \subset X,$$

is an isomorphism.

This lemma makes the natural connection between $Q_0(X)$ and $Q_0^+(X_+)$.

PROOF. Since the quadratic form $\langle Q_0 u, u\rangle$ is positive on X, we have that $Q_0 : X \to X^*$ is injective with dense $Q_0(X) \subset X^*$. As $\tilde{X}_{\leq 0}^* = \ker i_{X_+}^* \subset Q_0(X)$ due to (**B5**) and $X^* = \tilde{X}_{\leq 0}^* \oplus \tilde{X}_+^*$, we obtain that $\tilde{X}_+^* \cap Q_0(X)$ is dense in $\tilde{X}_+^*$ and $Q_0(X) = \tilde{X}_{\leq 0}^* \oplus \left(Q_0(X) \cap \tilde{X}_+^*\right)$. In the rest of the proof, we study $Q_0(X) \cap \tilde{X}_+^*$ and its associated properties.

Let $X_1 = Q_0^{-1}(\tilde{X}_{\leq 0}^*) \subset X$, which is a closed subspace. Since $\tilde{X}_{\leq 0}^* \subset Q_0(X)$ and Q_0 is injective, $Q_0 : X_1 \to \tilde{X}_{\leq 0}^*$ is bounded, injective, and surjective and thus an isomorphism. Let

$$\phi = (Q_0|_{X_1})^{-1} P_{\leq 0}^* : X_{\leq 0}^* \to X_1, \quad \phi_{\leq 0} = P_{\leq 0}\phi, \qquad \phi_+ = P_+\phi,$$

which are bounded operators. For any $f, g \in X_{\leq 0}^*$, since

$$\langle g, \phi_{\leq 0} f\rangle = \langle P_{\leq 0}^* g, \phi f\rangle = \langle Q_0 \phi g, \phi f\rangle,$$

we obtain that $\phi_{\leq 0} : X_{\leq 0}^* \to X_{\leq 0}$ is symmetric and $\langle f, \phi_{\leq 0} f\rangle > 0$ for any $0 \neq f \in X_{\leq 0}^*$. Therefore $\phi_{\leq 0}^{-1}$ is a densely defined closed operator satisfying $(\phi_{\leq 0}^{-1})^* = \phi_{\leq 0}^{-1} > 0$.

For any $f \in X_{\leq 0}^*$, let

$$\phi f = u_{\leq 0} + u_+, \quad u_+ = \phi_+ f, \quad u_{\leq 0} = \phi_{\leq 0} f,$$

then we have

$$Q_0^{\leq 0} u_{\leq 0} + Q_0^{\leq 0,+} u_+ = f, \quad Q_0^{+,\leq 0} u_{\leq 0} + Q_0^+ u_+ = 0.$$

It implies that $Q_0^{+,\leq 0} u_{\leq 0} \in Q_0^+(X_+)$ and $u_+ = -(Q_0^+)^{-1} Q_0^{+,\leq 0} u_{\leq 0}$. Therefore

$$\left(Q_0^{\leq 0} - Q_0^{\leq 0,+} (Q_0^+)^{-1} Q_0^{+,\leq 0}\right) u_{\leq 0} = f,$$

which implies that the closed positive symmetric operator $\phi_{\leq 0}^{-1}$ satisfies

$$0 < \phi_{\leq 0}^{-1} = Q_0^{\leq 0} - Q_0^{\leq 0,+} (Q_0^+)^{-1} Q_0^{+,\leq 0} \leq Q_0^{\leq 0}.$$

Here we also used $Q_0^{\leq 0,+} = (Q_0^{+,\leq 0})^*$ and the positivity of the symmetric closed operator $(Q_0^+)^{-1}$. Therefore $\phi_{\leq 0}$ is an isomorphism and

$$(Q_0^+)^{-1} Q_0^{+,\leq 0} = -\phi_+ \phi_{\leq 0}^{-1}$$

is bounded. The above inequality also implies the boundedness of $(Q_0^{\leq 0})^{-1} \leq \phi_{\leq 0}$.

On the one hand, for any $u \in X_+$, using $I = i_{X_{\le 0}} P_{\le 0} + i_{X_+} P_+$ we can write

$$P_+^* Q_0^+ u = Q_0 u - P_{\le 0}^* i_{X_{\le 0}}^* Q_0 u = Q_0 (I - \phi i_{X_{\le 0}}^* Q_0) u.$$

Therefore, $P_+^* Q_0^+(X_+) \subset Q_0(X) \cap \tilde{X}_+^*$ and

$$A \triangleq Q_0^{-1} P_+^* Q_0^+ = I - \phi i_{X_{\le 0}}^* Q_0 : X_+ \to X_2$$

is bounded, where $X_2 = Q_0^{-1}(\tilde{X}_+^*)$ is a closed subspace of X and $Q_0(X) \cap \tilde{X}_+^* = Q_0(X_2)$.

On the other hand, suppose $u = u_{\le 0} + u_+ \in X_2$, let $f = i_{X_+}^* Q_0 u \in X_+^*$ and $f_+ = P_+^* f = Q_0 u \in \tilde{X}_+^*$. We have

$$Q_0^{\le 0} u_{\le 0} + Q_0^{\le 0,+} u_+ = 0, \quad Q_0^{+,\le 0} u_{\le 0} + Q_0^+ u_+ = f,$$

and thus $u_{\le 0} = -(Q_0^{\le 0})^{-1} Q_0^{\le 0,+} u_+$. Substituting it into the second equation in the above, we obtain

$$f = \left(Q_0^+ - Q_0^{+,\le 0}(Q_0^{\le 0})^{-1} Q_0^{\le 0,+}\right) u_+ = Q_0^+ \tilde{u}_+,$$

where, from the above boundedness of $(Q_0^+)^{-1} Q_0^{+,\le 0}$,

$$\tilde{u}_+ = \left(I - (Q_0^+)^{-1} Q_0^{+,\le 0}(Q_0^{\le 0})^{-1} Q_0^{\le 0,+}\right) u_+ \in X_+.$$

It implies $f \in Q_0^+(X_+)$ and thus $f_+ \in P_+^* Q_0^+(X_+)$. Therefore

$$Q_0(X) \cap \tilde{X}_+^* \subset P_+^* Q_0^+(X_+).$$

Moreover, the above equality on $\tilde{u}_+$ also implies

$$A\left(I - (Q_0^+)^{-1} Q_0^{+,\le 0}(Q_0^{\le 0})^{-1} Q_0^{\le 0,+}\right) P_+ u = Q_0^{-1} P_+^* Q_0^+ \tilde{u}_+ = Q_0^{-1} f_+ = u.$$

Therefore we obtain

$$A^{-1} = \left(I - (Q_0^+)^{-1} Q_0^{+,\le 0}(Q_0^{\le 0})^{-1} Q_0^{\le 0,+}\right) P_+$$

is bounded and the proof of the lemma is complete. □

Construction of Y. As our main concern is that L_+ is not uniformly positive definite on X_+, we will actually work on the completion Y_+ of X_+ under the positive quadratic form $\langle L_+ \cdot, \cdot \rangle$.

We start with a resolution of identity to rewrite L_+ on X_+. From (**B3**), there exists $a > 0$ such that

$$\frac{1}{C}\|u\|^2 \le \|u\|_{L_+,a}^2 \le C\|u\|^2, \quad \forall u \in X_+, \tag{10.2}$$

for some $C > 0$, where, for $u, v \in X_+$, $\|u\|_{L_+,a}^2 = (u,u)_{L_+,a}$ and

$$(u,v)_{L_+,a} \triangleq \langle (L_+ + aQ_0^+)u, v\rangle = \langle (L + aQ_0)u, v\rangle.$$

For $u, v \in X_+$, let

$$\mathbb{L} = (L_+ + aQ_0^+)^{-1} L_+ : X_+ \to X_+,$$

which implies $(\mathbb{L}u, v)_{L_+,a} = \langle Lu, v\rangle$ and

$$\mathbb{D} = (Q_0^+)^{-1}(L_+ + aQ_0^+) : X_+ \supset D(\mathbb{D}) = (L_+ + aQ_0^+)^{-1} Q_0^+(X_+) \to X_+. \tag{10.3}$$

Clearly the Riesz representation $\mathbb{L}$ of L_+ with respect to the equivalent metric $(\cdot,\cdot)_{L_+,a}$ is a bounded symmetric linear operator. Since

$$\mathbb{D}^{-1} = (L_+ + aQ_0^+)^{-1} Q_0^+ = a^{-1}(I - \mathbb{L}) \tag{10.4}$$

is a bounded linear operator symmetric (and positive) with respect to $(\cdot,\cdot)_{L_+,a}$, $\mathbb{D}$ is self-adjoint with respect to $(\cdot,\cdot)_{L_+,a}$. In applications, if Q_1 is a uniformly positive elliptic operator and Q_0 corresponds to the L^2 duality, the operator $\mathbb{D}$ is basically a differential operator on X_+ of the same order as Q_1. The symmetric operator $\mathbb{L}$ admits a resolution of identity consisting of bounded projections $\Pi_\lambda : X_+ \to X_+$, $\lambda \in [0,1]$ where

(1) $\lim_{\lambda\to\lambda_0+} \Pi_\lambda u = \Pi_{\lambda_0} u$, for all $\lambda_0 \in [0,1)$ and $u \in X_+$;
(2) $\Pi_{\lambda_1}\Pi_{\lambda_2} = \Pi_{\min\{\lambda_1,\lambda_2\}}$, for all $\lambda_{1,2} \in [0,1]$;
(3) $\langle (L_+ + aQ_0^+)\Pi_\lambda u_1, u_2\rangle = \langle (L_+ + aQ_0^+)u_1, \Pi_\lambda u_2\rangle$ for any $u \in X_+$ and $\lambda \in [0,1]$;
(4) $u = \int_0^1 d\Pi_\lambda u$, $\mathbb{L}u = \int_0^1 \lambda\, d\Pi_\lambda u$, for any $u \in X_+$.

Here $\Pi_1 = I$ and $\Pi_0 = 0$ since L_+ is bounded and $0 < L_+ < L_+ + aQ_0^+$ as a quadratic form. Using this resolution of identity, we have the representations of L_+ and $\|\cdot\|_{L_+,a}$

$$\langle L_+u, v\rangle = \int_0^1 \lambda\, d(\Pi_\lambda u, v)_{L_+,a}, \quad \|u\|^2_{L_+,a} = \int_0^1 d\,\|\Pi_\lambda u\|^2_{L_+,a}, \quad u, v \in X_+.$$

Let $(Y_+, \|\cdot\|_{L_+})$ be the Hilbert space of the completion of X_+ with respect to the inner product

$$(u,v)_{L_+} = (\mathbb{L}u, v)_{L_+,a} = \langle L_+u, v\rangle = \langle Lu, v\rangle = \int_0^1 \lambda\, d(\Pi_\lambda u, v)_{L_+,a}, \quad u, v \in X_+.$$

Therefore X_+ is densely embedded into Y_+ through the embedding i_{X_+}. Using the above spectral integral representation of $\mathbb{L}$, one may extend Π_λ to be bounded linear projections on Y orthogonal with respect to $(\cdot,\cdot)_{L_+}$ as well, satisfying $|\Pi_\lambda|_Y \le 1$. Moreover, for $\lambda \in (0,1]$, $(I - \Pi_\lambda)Y_+ \subset X_+$ and

$$\begin{aligned} &\forall\, u \in X_+, \ \|\Pi_\lambda u\|_{L_+} \le \lambda \|\Pi_\lambda u\|_{L_+,a}, \\ &\forall\, u \in Y_+, \ \lambda\|(I-\Pi_\lambda)u\|_{L_+,a} \le \|(I-\Pi_\lambda)u\|_{L_+} \le \|(I-\Pi_\lambda)u\|_{L_+,a}, \end{aligned} \tag{10.5}$$

where $I - \Pi_\lambda = \int_{(\lambda,1]} d\Pi_\lambda$ was used.

As Y_+ is defined as the completion of X_+ with respect to the metric $(\mathbb{L}u, u)_{L_+,a}$, elements in Y_+ are defined via Cauchy sequences in X_+ with respect to this metric. This is rather inconvenient technically. Instead we give an integral representation of elements in Y_+ and some linear quantities on Y_+ using Π_λ and the following lemma.

LEMMA 10.2. *$\lim_{\lambda\to 0+} \|\Pi_\lambda u\|_{L_+} = 0$ for any $u \in Y_+$.*

PROOF. For any $\epsilon > 0$, there exists $v \in X_+$ such that $\|u - v\|_{L_+} < \frac{\epsilon}{2}$. Since $\lim_{\lambda\to 0+} \Pi_\lambda v = \Pi_0 v = 0$ in X_+, there exists $\lambda_0 > 0$ such that $\|\Pi_\lambda v\|_{L_+,a} < \frac{\epsilon}{2}$ for any $\lambda \in (0,\lambda_0)$. Therefore, for any $\lambda \in (0,\lambda_0)$,

$$\|\Pi_\lambda u\|_{L_+} \le \|\Pi_\lambda(u-v)\|_{L_+} + \|\Pi_\lambda v\|_{L_+} \le \|u-v\|_{L_+} + \lambda\|\Pi_\lambda v\|_{L_+,a} \le \epsilon.$$

The lemma is proved. □

COROLLARY 10.1. *For any $u, v \in Y_+$, we have*

$$
\begin{aligned}
&u = \int_0^1 d\Pi_\lambda u = -\int_0^1 d(I - \Pi_\lambda)u = -\lim_{\lambda\to 0+}\int_\lambda^1 d(I-\Pi_\lambda)u,\\
&\mathbb{L}u = -\int_0^1 \lambda d(I-\Pi_\lambda)u, \ \|u\|_{L_+}^2 = -\int_0^1 \lambda d\|(I-\Pi_\lambda)u\|_{L_+,a}^2,\\
&\langle L_+u, v\rangle = -\int_0^1 \lambda d\big((I-\Pi_\lambda)u, v\big)_{L_+,a} = \lim_{\lambda\to 0+}\langle L_+(I-\Pi_\lambda)u, (I-\Pi_\lambda)v\rangle,\\
&Y_+^* = \{f = (L_+ + aQ_0^+)u \mid u \in X_+,\\
&\qquad \|f\|_{Y_+^*}^2 = -\int_0^1 \lambda^{-1} d\|(I-\Pi_\lambda)u\|_{L_+,a}^2 < \infty\} \subset X_+^*.
\end{aligned}
$$

Here the first integral converges in the $\|\cdot\|_{L_+}$ norm and the minus signs are due to the non-increasing monotonicity of $\|(I-\Pi_\lambda)u\|_{L_+,a}^2$. With $(I-\Pi_\lambda)u \in X_+$ for $\lambda \in (0,1]$, these integral representations are more convenient than the Cauchy sequence representations of elements in Y_+. In particular, for $f = (L_+ + aQ_0^+)u \in Y_+^*$ and $v \in Y_+$,

$$
\begin{aligned}
\langle f, v\rangle = &-\int_0^1 d\big((I-\Pi_\lambda)u, v\big)_{L_+,a} = \lim_{\lambda\to 0+}\langle (L_+ + aQ_0^+)(I-\Pi_\lambda)u, (I-\Pi_\lambda)v\rangle\\
&\le \|f\|_{Y_+^*}\|v\|_{L_+}.
\end{aligned}
$$

Let

$$
Y = X_{\le 0} \oplus Y_+, \quad (u,v)_Y = (P_{\le 0}u, P_{\le 0}v) + \big((I-P_{\le 0})u, (I-P_{\le 0})v\big)_{L_+}, \tag{10.6}
$$

where, with slight abuse of notations, $P_{\le 0} : Y \to X_{\le 0}$ represents the projection operator with kernel Y_+. Clearly X is densely embedded into Y and let i_X denote the embedding.

The dual space Y^* is densely embedded into X^* through i_X^* and thus can be viewed as a dense subspace of X^*. It is straightforward to see that $i_X^* Y^* = \tilde{X}_{\le 0}^* \oplus \tilde{Y}_+^*$,

$$
\langle f, v\rangle = \langle g, u\rangle = 0, \ \forall\, u \in X_{\le 0}, \ v \in Y_+, \ f \in \tilde{X}_{\le 0}^*, \ g \in \tilde{Y}_+^*
$$

and

$$
\begin{aligned}
\tilde{Y}_+^* = \tilde{X}_+^* \cap i_X^*(Y^*) = P_+^*\{&f = (L_+ + aQ_0^+)u \mid u \in X_+,\\
&\|f\|_{Y_+^*}^2 = -\int_0^1 \lambda^{-1} d\|(I-\Pi_\lambda)u\|_{L_+,a}^2 < \infty\} \subset X^*.
\end{aligned} \tag{10.7}
$$

Operator L is naturally extended as a bounded symmetric linear operator $L_Y : Y \to Y^*$ by

$$
\langle L_Y u, v\rangle = \langle L_{\le 0}P_{\le 0}u, P_{\le 0}v\rangle + \langle L_+(I-P_{\le 0})u, (I-P_{\le 0})v\rangle, \tag{10.8}
$$

where L_+ on Y_+ is computed by the formula given in Corollary 10.1.

From assumption (**B4**) on the L-orthogonality of the decomposition

$$
X = X_- \oplus \ker L \oplus X_+ = X_{\le 0} \oplus X_+
$$

and Corollary 10.1, the operator L_Y defined in the above satisfies (**H2**) on Y with $\delta = 1$ in (**H2.b**).

Operator J_Y. We define $J_Y : Y^* \supset D(J_Y) \to Y$ essentially as the restriction of J on Y^*, namely,

$$J_Y \triangleq i_X \mathbb{J} Q_0^{-1} i_X^* : Y^* \supset D(J_Y) \to Y, \quad D(J_Y) = (i_X^*)^{-1} Q_0(X) \subset Y^*, \tag{10.9}$$

where we recall that $i_X : X \to Y$ is the embedding. Assumption (**H3**) is satisfied due to (**B5**) and (10.1). Therefore, to complete the proof of Theorem 2.7, it suffice to prove $J_Y^* = -J_Y$.

LEMMA 10.3. *It holds that $i_X^* D(J_Y)$ is dense in X^* and*

$$i_X^* D(J_Y) = Q_0(X) \cap i_X^* Y^* = \tilde{X}_{\leq 0}^* \oplus P_+^*(L_+ + aQ_0^+) X_{1+},$$

where

$$X_{1+} = \{u \in X_+ \mid \int_0^1 \frac{-1}{\lambda(1-\lambda)^2} d\|(I - \Pi_\lambda)u\|^2 < \infty\} \subset X_+.$$

PROOF. From $i_X^* Y^* = \tilde{X}_{\leq 0}^* \oplus \tilde{Y}_+^*$ and (**B5**), we can decompose

$$i_X^* D(J_Y) = Q_0(X) \cap i_X^* Y^* = \tilde{X}_{\leq 0}^* \oplus \big(\tilde{Y}_+^* \cap Q_0(X)\big).$$

As $\tilde{Y}_+^* \subset \tilde{X}_+^*$, we obtain from Lemma 10.1

$$i_X^* D(J_Y) = \tilde{X}_{\leq 0}^* \oplus \big(\tilde{Y}_+^* \cap P_+^* Q_0^+(X_+)\big). \tag{10.10}$$

Recall (10.3) and we have $(L_+ + aQ_0^+)u \in Q_0^+(X_+)$, $u \in X_+$, if and only if $u \in D(\mathbb{D})$, which is equivalent to $u \in (I - \mathbb{L})(X_+)$ according to (10.4). Therefore we obtain that

$$(L_+ + aQ_0^+)u \in Q_0^+(X_+), \ u \in X_+,$$

if and only if

$$-\int_0^1 (1-\lambda)^{-2} d\|(I - \Pi_\lambda)u\|_{L_+,a}^2 < \infty,$$

which can be seen from

$$\mathbb{D}u = (Q_0^+)^{-1}(L_+ + aQ_0^+)u = -\frac{1}{a}\int_0^1 (1-\lambda)^{-1} d(I - \Pi_\lambda)u. \tag{10.11}$$

The lemma follows immediately from this property and the characterization (10.7) of $\tilde{Y}_+^*$. □

To prove $J_Y^* = -J_Y$, suppose $f \in D(J_Y^*)$ and $u = J_Y^* f$, namely, $f \in Y^*$ and $u \in Y$ satisfies

$$\langle f, J_Y g\rangle = \langle g, u\rangle, \quad \forall g \in D(J_Y). \tag{10.12}$$

Firstly, for any $\epsilon \in (0, \frac{1}{2})$, take

$$g = -(i_X^*)^{-1} P_+^*(L_+ + aQ_0^+) \int_\epsilon^{\frac{1}{2}} d(I - \Pi_\lambda) P_+ u \in D(J_Y) \cap (i_X^*)^{-1} \tilde{X}_+^*,$$

where Lemma 10.3 is used. Equalities (10.12) and (10.11) imply

$$-\int_\epsilon^{\frac{1}{2}} d\|(I - \Pi_\lambda)P_+ u\|_{L_+,a}^2 = \langle g, u\rangle = \langle f, J_Y g\rangle = \langle \mathbb{J}^* i_X^* f, Q_0^{-1} i_X^* g\rangle$$

$$= -\langle \mathbb{J}^* i_X^* f, A\mathbb{D} \int_\epsilon^{\frac{1}{2}} d(I - \Pi_\lambda) P_+ u\rangle = -\frac{1}{a}\langle \mathbb{J}^* i_X^* f, A \int_\epsilon^{\frac{1}{2}} (1-\lambda)^{-1} d(I - \Pi_\lambda) P_+ u\rangle,$$

where A is defined in Lemma 10.1 and proved to be bounded. Since $\lambda \in (0, \frac{1}{2}]$, there exists $C > 0$ such that

$$-\int_{\epsilon}^{\frac{1}{2}} d\|(I - \Pi_\lambda)P_+ u\|_{L_+,a}^2 \leq C\|f\|_{Y^*}^2, \quad \forall \epsilon \in (0, \frac{1}{2}).$$

Therefore we obtain $u \in X$, or more precisely,

$$u = i_X \tilde{u}, \quad \tilde{u} \in X.$$

For any $g \in D(J_Y)$, from (**B1-2**) we can compute

$$\begin{aligned}
&\langle i_X^* f, \mathbb{J} Q_0^{-1} i_X^* g \rangle = \langle f, J_Y g \rangle = \langle g, u \rangle = \langle g, i_X \tilde{u} \rangle \\
=&\langle Q_0 Q_0^{-1} i_X^* g, \tilde{u} \rangle = \langle Q_0 \mathbb{J} Q_0^{-1} i_X^* g, \mathbb{J}\tilde{u} \rangle = \langle Q_0 \mathbb{J} \tilde{u}, \mathbb{J} Q_0^{-1} i_X^* g \rangle.
\end{aligned}$$

Since $\mathbb{J}$ is assumed to be isomorphic in (**B2**), Q_0^{-1} is surjective, and $i_X^* D(J_Y)$ is dense in X^* (Lemma 10.3), we obtain

$$i_X^* f = Q_0 \mathbb{J} \tilde{u} \in Q_0(X).$$

Thus it follows from (**B2**) that

$$J_Y f = i_X \mathbb{J}^2 \tilde{u} = -u,$$

which implies $J_Y^* \subset -J_Y$. Again from (**B2**), it is easy to see that J_Y is symmetric, namely, $J_Y \subset -J_Y^*$. Therefore we complete the proof of Theorem 2.7.

CHAPTER 11

Hamiltonian PDE models

In this chapter, based on the above general theory, we study the stability issues of examples of Hamiltonian PDEs including several dispersive wave models, the 2D Euler equation for inviscid flows and a 2D nonlinear Schrödinger equations with nonzero conditions at infinity.

First, in Sections 11.1 to 11.3, we study the stability/instability of traveling solitary and periodic wave solutions of several classes of equations modeling weakly nonlinear dispersive long waves. They include BBM, KDV, and good Boussinesq type equations. These equations respectively have the forms:

1. BBM type

$$\partial_t u + \partial_x u + \partial_x f(u) + \partial_t \mathcal{M} u = 0; \tag{11.1}$$

2. KDV type

$$\partial_t u + \partial_x f(u) - \partial_x \mathcal{M} u = 0; \tag{11.2}$$

3. good Boussinesq (gBou) type

$$\partial_t^2 u - \partial_x^2 u + \partial_x^2 f(u) - \partial_x^2 \mathcal{M} u = 0. \tag{11.3}$$

We follow the notations in [**52**]. Here, the pseudo-differential operator $\mathcal{M}$ is defined as

$$\widehat{\mathcal{M}g}(\xi) = \alpha(\xi)\widehat{g}(\xi),$$

where $\hat{g}$ is the Fourier transformation of g. We assume: i) f is C^1 with $f(0) = f'(0) = 0$, and $f(u)/u \to \infty$. ii) $a|\xi|^m \leq \alpha(\xi) \leq b|\xi|^m$ for large ξ, where $m > 0$ and $a, b > 0$. If $f(u) = u^2$ and $\mathcal{M} = -\partial_x^2$, the above equations recover the original BBM, KDV, and good Boussinesq equations, which have been used to model the propagation of water waves of long wavelengths and small amplitude.

11.1. Stability of Solitary waves of Long wave models

Consider the equations (11.1)-(11.3) with $(x,t) \in \mathbf{R} \times \mathbf{R}$. Up to a shift of a constant of the wave speed/symbol $\alpha(\xi)$, we can assume that $\sigma_{ess}(\mathcal{M}) \subset [0, \infty)$. Each of the equations (11.1)-(11.3) admits solitary-wave solutions of the form $u(x,t) = u_c(x - ct)$ for $c > 1, c > 0, c^2 < 1$ respectively, where $u_c(x) \to 0$ as $|x| \to \infty$. They satisfy the equations

$$\mathcal{M} u_c + \left(1 - \frac{1}{c}\right) u_c - \frac{1}{c} f(u_c) = 0, \text{ (BBM)} \tag{11.4}$$

$$\mathcal{M} u_c + c u_c - f(u_c) = 0, \text{ (KDV)}$$

and

$$\mathcal{M} u_c + \left(1 - c^2\right) u_c - f(u_c) = 0, \text{ (gBou)}$$

respectively. We refer to the introduction of [**52**] and the book [**2**] for the literature on the existence of such solitary waves. Before stating the results, we introduce some notations. For BBM type equations (11.1), define the operator

$$\mathcal{L}_0 = \mathcal{M} + \left(1 - \frac{1}{c}\right) - \frac{1}{c} f'(u_c) : H^m \to L^2, \tag{11.5}$$

and the momentum

$$P(c) = \frac{1}{2} \int u_c (\mathcal{M}+1) u_c. \tag{11.6}$$

For KDV type equations (11.2), define

$$\mathcal{L}_0 := \mathcal{M} + c - f'(u_c), \quad P(c) = \frac{1}{2} \int u_c^2. \tag{11.7}$$

For good Boussinesq type equations (11.3), define

$$\mathcal{L}_0 := \mathcal{M} + 1 - c^2 - f'(u_c), \ P(c) = -c \int u_c^2. \tag{11.8}$$

Denote by $n^-(\mathcal{L}_0)$ the number (counting multiplicity) of negative eigenvalues of the operators $\mathcal{L}_0$.

The linearizations of (11.1)-(11.3) in the traveling frame $(x - ct, t)$ are

$$(\partial_t - c\partial_x)(u + \mathcal{M}u) + \partial_x(u + f'(u_c) u) = 0, \text{ (BBM)} \tag{11.9}$$

$$(\partial_t - c\partial_x) u + \partial_x (f'(u_c) u - \mathcal{M}u) = 0, \text{ (KDV)} \tag{11.10}$$

and

$$(\partial_t - c\partial_x)^2 u - \partial_x^2 (u - f'(u_c) u + \mathcal{M}u) = 0, \text{ (gBou)} \tag{11.11}$$

respectively. We consider the Hamiltonian structures of these equations.

For BBM type equations, (11.9) can be written as $\partial_t u = JLu$, where $J = c\partial_x (1 + \mathcal{M})^{-1}$ and $L = \mathcal{L}_0$ is defined in (11.5). By differentiating (11.4) in x and c, we have $\mathcal{L}_0 u_{c,x} = 0$ and

$$\mathcal{L}_0 \partial_c u_c = -\frac{1}{c} (1 + \mathcal{M}) u_c$$

which implies that $J\mathcal{L}_0 \partial_c u_c = -u_{c,x}$ and $\langle \mathcal{L}_0 \partial_c u_c, \partial_c u_c \rangle = -\frac{1}{c} dP/dc$.

For KDV type equations, (11.10) is written as $\partial_t u = J\mathcal{L}_0 u$, where $J = \partial_x$ and $L = \mathcal{L}_0$ is defined in (11.7). Similarly, $\mathcal{L}_0 u_{c,x} = 0$, $\mathcal{L}_0 \partial_c u_c = -u_c$, and

$$J\mathcal{L}_0 \partial_c u_c = -u_{c,x}, \ \langle \mathcal{L}_0 \partial_c u_c, \partial_c u_c \rangle = -dP/dc. \tag{11.12}$$

For good Boussinesq type equations, we write (11.11) as a first order system. Let $(\partial_t - c\partial_x) u = v_x$, then

$$(\partial_t - c\partial_x) v = \partial_x (\mathcal{M} + 1 - f'(u_c)) u = \partial_x \left(\mathcal{L}_0 + c^2\right) u.$$

Thus

$$\partial_t \begin{pmatrix} u \\ v \end{pmatrix} = JL \begin{pmatrix} u \\ v \end{pmatrix},$$

with

$$J = \begin{pmatrix} 0 & \partial_x \\ \partial_x & 0 \end{pmatrix}, \quad L = \begin{pmatrix} \mathcal{L}_0 + c^2 & c \\ c & 1 \end{pmatrix}. \tag{11.13}$$

We have

$$\ker L = \{(u, -cu) \mid u \in \ker \mathcal{L}_0\}.$$

Since

$$\left\langle L\begin{pmatrix} u \\ v \end{pmatrix}, \begin{pmatrix} u \\ v \end{pmatrix}\right\rangle = \langle \mathcal{L}_0 u, u\rangle + \int (v+cu)^2, \tag{11.14}$$

so $n^-(L) = n^-(\mathcal{L}_0)$. Similarly as in BBM and KDV types, we have

$$L\begin{pmatrix} u_{c,x} \\ -cu_{c,x} \end{pmatrix} = 0,\ L\begin{pmatrix} -\partial_c u_c \\ c\partial_c u_c + u_c \end{pmatrix} = \begin{pmatrix} -cu_c \\ u_c \end{pmatrix},$$

$$JL\begin{pmatrix} -\partial_c u_c \\ c\partial_c u_c + u_c \end{pmatrix} = \begin{pmatrix} u_{c,x} \\ -cu_{c,x} \end{pmatrix},$$

and

$$\left\langle L\begin{pmatrix} -\partial_c u_c \\ c\partial_c u_c + u_c \end{pmatrix}, \begin{pmatrix} -\partial_c u_c \\ c\partial_c u_c + u_c \end{pmatrix}\right\rangle = -dP/dc,$$

where P is defined in (11.8). For all three cases, we have $\sigma_{ess}(\mathcal{L}_0) \subset [\delta_0, \infty)$ for some $\delta_0 > 0$. So the quadratic form $\langle \mathcal{L}_0 \cdot, \cdot\rangle$ is positive definite on $H^{\frac{m}{2}}$ in a finite codimensional space. This along with Remark 2.3 shows that the quadratic form $\langle \mathcal{L}_0 \cdot, \cdot\rangle$ in the Hamiltonian formulation of BBM and KDV type equations satisfies the assumption (**H1-3**) in the general framework with $X = H^{\frac{m}{2}}$. By (11.14), the quadratic form $\langle L\cdot, \cdot\rangle$ in the Hamiltonian formulation (11.13) of good Boussinesq type equations also satisfies (**H1-3**) in the space $(u, \partial_t u) \in X = H^{\frac{m}{2}} \times L^2$. Thus by Theorems 2.2, 2.3, and Corollary 2.2, we get the following results.

THEOREM 11.1. *Consider the linearized equations (11.9)-(11.11) at solitary waves $u_c(x-ct)$ of equations (11.1)-(11.3). Then: (i) The following index formula holds*

$$k_r + 2k_c + 2k_i^{\leq 0} + k_0^{\leq 0} = n^-(\mathcal{L}_0). \tag{11.15}$$

(ii) The linear exponential trichotomy holds in the space $H^{\frac{m}{2}}$ for the linearized equations (11.9) and (11.10), and in $H^{\frac{m}{2}} \times L^2$ for (11.11).
(iii) When $dP/dc \geq 0$, we have $k_0^{\leq 0} \geq 1$. Moreover, if $\ker \mathcal{L}_0 = span\{u_{c,x}\}$, then

$$k_0^{\leq 0} = \begin{cases} 1 & \text{if } dP/dc > 0 \\ 0 & \text{if } dP/dc < 0 \end{cases}.$$

COROLLARY 11.1. *(i) When $dP/dc \geq 0$ and $n^-(\mathcal{L}_0) \leq 1$, the spectral stability holds true.*
(ii) If $\ker \mathcal{L}_0 = span\{u_{c,x}\}$, then there is linear instability when $n^-(\mathcal{L}_0)$ is even and $dP/dc > 0$ or $n^-(\mathcal{L}_0)$ is odd and $dP/dc < 0$.

In particular, when $\mathcal{M} = -\partial_x^2$, by the fact that $u_{c,x}$ changes sign exactly once and the Sturm-Liouville theory, we have $\ker \mathcal{L}_0 = span\{u_{c,x}\}$ and $n^-(\mathcal{L}_0) = 1$. Thus, we have

COROLLARY 11.2. *When $\mathcal{M} = -\partial_x^2$ and $dP/dc < 0$, for the linearized equations (11.1)-(11.3), we have $k_r = 1$ and $k_c = k_i^- = 0$. In particular, on the center space E^c as given in Theorem 2.2, we have*

$$\langle L\cdot, \cdot\rangle|_{E^c \cap \{u_{c,x}\}^{\perp}} \geq \delta_0 > 0. \tag{11.16}$$

The stability and instability of solitary waves of dispersive models had been studied a lot in the literature. Assume $\ker \mathcal{L}_0 = \{u_{c,x}\}$, $n^-(\mathcal{L}_0) = 1$, then when $dP/dc > 0$, the orbital stability of traveling solitary waves was proved (e.g. [**5**] [**29**] [**10**]) by using the method of Lyapunov functionals. When $dP/dc < 0$, the nonlinear instability was proved in [**9**] [**70**] for generalized BBM and KDV equations, and in [**58**] for good Boussinesq equation. The instability proof in these papers was by contradiction argument which bypassed the linearized equation. The existence of unstable eigenvalues when $dP/dc > 0$ was proved in [**63**] for KDV and BBM equations. In [**52**], an instability criterion as in Corollary 11.1 (ii) was proved for KDV and BBM type equations. In [**63**] and [**52**], an instability criterion was also given for the regularized Boussinesq equation which takes an indefinite Hamiltonian form (i.e. $n^-(L) = \infty$) and is therefore not included in the framework of this paper. Recently, in [**46**] and [**65**], an instability index theorem similar to (11.15) was given for KDV and BBM type equations under the assumption that $\dim\ker\mathcal{L}_0 = 1$ and $dP/dc \neq 0$. The proof of [**46**] [**65**] was by using ad-hoc arguments to transform the eigenvalue problem $\partial_x\mathcal{L}_0 u = \lambda u$ to another Hamiltonian form with a symplectic operator which has a bounded inverse. The linear instability of solitary waves of good Boussinesq equation ($\mathcal{M} = -\partial_x^2$) was studied in [**1**] by Evans function and in [**68**] by using quadratic operator pencils. The index formula (11.15) for the good Boussinesq type equations appears to be new.

Besides giving a more unified and general index formula for linear instability, Theorem 11.1 also gives the exponential trichotomy for $e^{tJ\mathcal{L}_0}$, which is an important step for constructing invariant manifolds (stable, unstable and center) near the translation orbit of u_c. Moreover, when $\ker\mathcal{L}_0 = span\{u_{c,x}\}$ and $n^-(\mathcal{L}_0) = 1$, there exists a pair of stable and unstable eigenvalues and $\mathcal{L}_0$ is positive on the codimension two center space modulo the translation kernel. This positivity property has an important implication for the center manifolds once constructed. For example, in [**37**], the invariant (stable, unstable and center) manifolds were constructed near the orbit of unstable solitary waves of generalized KDV equation in the energy space. More precisely, there exist 1-d stable and unstable manifolds and co-dimension two center manifold near the translation orbits of unstable solitary waves. These invariant manifolds give a complete description of local dynamics near unstable traveling wave orbit. The positivity estimate (11.16) on the center subspace implies that on the codimension two center manifold, the solitary wave u_c is orbitally stable, which in turn also leads to the local uniqueness of the center manifold. Any initial data not lying on the center manifold will leave the orbit neighborhood of unstable traveling waves exponentially fast.

11.2. Stability of periodic traveling waves

Consider the equations (11.1)-(11.3) in the periodic case. For convenience we assume the period is 2π, that is, $(x,t) \in \mathbf{S}^1 \times \mathbf{R}$. A periodic traveling wave is of the form $u(x,t) = u_{c,a}(x - ct)$, where $u_{c,a}$ satisfies the equations

$$\mathcal{M}u_{c,a} + \left(1 - \frac{1}{c}\right)u_{c,a} - \frac{1}{c}f(u_{c,a}) = a, \text{ (BBM)} \tag{11.17}$$

$$\mathcal{M}u_{c,a} + cu_{c,a} - f(u_{c,a}) = a, \text{ (KDV)} \tag{11.18}$$

and

$$\mathcal{M}u_{c,a} + \left(1 - c^2\right)u_{c,a} - f(u_{c,a}) = a, \text{ (gBou)} \tag{11.19}$$

for some constant a. In this section, we consider the perturbations of the same period 2π (i.e. co-periodic perturbations) and leave the case of different periods to the next section. The linearized equations in the traveling frame $(x - ct, t)$ near traveling waves $u_{c,a}$ take the same form (11.9)-(11.11). Their Hamiltonian structures are formally the same as in the case of solitary waves. However, the operator J has rather different spectral properties in the periodic case. More precisely, for solitary waves the symplectic operators J, which is $c\partial_x (1+\mathcal{M})^{-1}$ for BBM, ∂_x for KDV and $\begin{pmatrix} 0 & \partial_x \\ \partial_x & 0 \end{pmatrix}$ for good Boussinesq, has no kernel in $L^2(\mathbf{R})$. But for the periodic case, J has nontrivial kernel in X^*. Indeed, $\ker J = span\{1\}$ for BBM and KDV, and

$$\ker J = span\{\vec{e}_1, \vec{e}_2\} = span\left\{\begin{pmatrix} 1 \\ 0 \end{pmatrix}, \begin{pmatrix} 0 \\ 1 \end{pmatrix}\right\}$$

for good Boussinesq. This degeneracy of J leads to the extra free parameter a in traveling waves.

We now discuss the consequential changes in the index formula induced by the nontrivial kernel of J. For BBM type equations, define the operator $\mathcal{L}_0 : H^m(\mathbf{S}^1) \to L^2(\mathbf{S}^1)$ and the momentum P as in (11.5) and (11.6). Differentiating (11.17) we obtain

$$R(\mathcal{L}_0) \ni \mathcal{L}_0 \partial_a u_{c,a} = 1.$$

Let

$$U_{c,a} = \partial_a u_{c,a},\ d_1 = \int_{\mathbf{S}^1} U_{c,a}\, dx,\ N = \int_{\mathbf{S}^1} u_{c,a} dx \text{ (total mass)}. \tag{11.20}$$

We have $\mathcal{L}_0 \partial_x u_{c,a} = 0$ and from differentiating (11.17)

$$\mathcal{L}_0 \partial_c u_{c,a} = -\frac{1}{c}(1+\mathcal{M}) u_{c,a} + \frac{a}{c},$$

and thus $J\mathcal{L}_0 \partial_c u_{c,a} = -\partial_x u_{c,a}$. Denote

$$D = \begin{pmatrix} \langle \mathcal{L}_0 U_{c,a}, U_{c,a} \rangle & \langle \mathcal{L}_0 U_{c,a}, \partial_c u_{c,a} \rangle \\ \langle \mathcal{L}_0 U_{c,a}, \partial_c u_{c,a} \rangle & \langle \mathcal{L}_0 \partial_c u_{c,a}, \partial_c u_{c,a} \rangle \end{pmatrix} = \begin{pmatrix} d_1 & N'(c) \\ N'(c) & -\frac{1}{c} dP/dc + \frac{a}{c} N'(c) \end{pmatrix} \tag{11.21}$$

that is, the matrix for $\langle L\cdot, \cdot \rangle$ on $span\{U_{c,a}, \partial_c u_{c,a}\} \subset g\ker(J\mathcal{L}_0)$. Denote $n^{\leq 0}(D)$ to be the number of non-positive eigenvalues of D.

For KDV type equations, similarly, $\mathcal{L}_0 \partial_x u_{c,a} = 0$,

$$\mathcal{L}_0 \partial_c u_{c,a} = -u_{c,a},\ J\mathcal{L}_0 \partial_c u_{c,a} = -\partial_x u_{c,a},\ \mathcal{L}_0 \partial_a u_{c,a} = 1,$$

and we define $U_{c,a}, d_1, N, D, n^{\leq 0}(D)$ etc. as in (11.20) and (11.21).

For good Boussinesq type equations, still define $U_{c,a}, d_1, N$ as in (11.20). Let

$$\vec{U}_1 = \begin{pmatrix} U_{c,a} \\ -cU_{c,a} \end{pmatrix},\ \vec{U}_2 = \begin{pmatrix} -cU_{c,a} \\ 1 + c^2 U_{c,a} \end{pmatrix}, \vec{U}_3 = \begin{pmatrix} -\partial_c u_{c,a} \\ c\partial_c u_{c,a} + u_{c,a} \end{pmatrix}, \tag{11.22}$$

then

$$L\vec{U}_1 = \vec{e}_1,\ \ L\vec{U}_2 = \vec{e}_2,\ L\vec{U}_3 = \begin{pmatrix} -cu_{c,a} \\ u_{c,a} \end{pmatrix}.$$

Define the matrix D of $\langle L\cdot,\cdot\rangle$ on the space spanned by $\left\{\vec{U}_1,\ \vec{U}_2,\vec{U}_3\right\}$, that is,

$$D=\begin{pmatrix}\left\langle L\vec{U}_1,\vec{U}_1\right\rangle & \left\langle L\vec{U}_1,\vec{U}_2\right\rangle & \left\langle L\vec{U}_1,\vec{U}_3\right\rangle\\ \left\langle L\vec{U}_1,\vec{U}_2\right\rangle & \left\langle L\vec{U}_2,\vec{U}_2\right\rangle & \left\langle L\vec{U}_3,\vec{U}_2\right\rangle\\ \left\langle L\vec{U}_1,\vec{U}_3\right\rangle & \left\langle L\vec{U}_3,\vec{U}_2\right\rangle & \left\langle L\vec{U}_3,\vec{U}_3\right\rangle\end{pmatrix} \tag{11.23}$$
$$=\begin{pmatrix} d_1 & -cd_1 & -N'(c)\\ -cd_1 & \int_{\mathbf{S}^1}\left(1+c^2U_{c,a}\right)dx & cN'(c)+N(c)\\ -N'(c) & cN'(c)+N(c) & -P'(c)\end{pmatrix}.$$

Again $n^{\leq 0}(D)$ denotes the number of non-positive eigenvalues of D.

Since in the periodic case, the operator $\mathcal{L}_0$ has only discrete spectrum which tends to $+\infty$, it is easy to verify that assumptions (**H1-3**) are satisfied in $X=H^{\frac{m}{2}}$ for BBM and KdV type equations and $X=H^{\frac{m}{2}}\times L^2$ for good Boussinesq type equations. Thus similar to Theorem 11.1, we have

THEOREM 11.2. *Consider the linearized equations (11.9)-(11.11) near periodic waves $u_c(x-ct)$ of equations (11.1)-(11.3). Then: (i) the following index formula holds*

$$k_r+2k_c+2k_i^-+k_0^-=n^-\left(\mathcal{L}_0\right).$$

(ii) the linear exponential trichotomy is true in the space $H^{\frac{m}{2}}\left(\mathbf{S}^1\right)$ for the linearized equations (11.9) and (11.10), and in $H^{\frac{m}{2}}\left(\mathbf{S}^1\right)\times L^2\left(\mathbf{S}^1\right)$ for (11.11). (iii) $k_0^-\geq n^{\leq 0}(D)$, the number of non-positive eigenvalues of the matrix D defined in (11.21), (11.22) and (11.23). Moreover, when $\ker\mathcal{L}_0=\{\partial_x u_{c,a}\}$ and D is nonsingular, $k_0^-=n^-(D)$ (the number of negative eigenvalues of D) and we have

$$k_r+2k_c+2k_i^-=n^-\left(\mathcal{L}_0\right)-n^-(D). \tag{11.24}$$

As corollaries, we have from Proposition 2.7 and Remark 2.14 the following linear stability/instability conditions.

COROLLARY 11.3. *(i) If $n^{\leq 0}(D)\geq n^-\left(\mathcal{L}_0\right)$, then the spectral stability holds.*

(ii) If $\ker\mathcal{L}_0=span\left\{\partial_x u_{c,a}\right\}$, D is nonsingular and $n^-\left(\mathcal{L}_0\right)-n^-(D)$ is odd, then there is linear instability.

When $n^-\left(\mathcal{L}_0\right)=n^-(D)$, nonlinear orbital stability holds for (11.1)-(11.3) as well. More precisely, we have

PROPOSITION 11.1. *When $\ker\mathcal{L}_0=span\left\{\partial_x u_{c,a}\right\}$, D is nonsingular and $n^-\left(\mathcal{L}_0\right)=n^-(D)$, then there is orbital stability in X of the traveling waves $u_{c,a}(x-ct)$ of equations (11.1)-(11.3) for perturbations of the same period.*

PROOF. Here we sketch the proof based on the standard Lyapunov functional method (e.g. [**29**], [**30**]). Consider the KDV type equation (11.2). It has three invariants: (1) energy $E(u)=\int\left[\frac{1}{2}u\mathcal{M}u-F(u)\right]dx$, with $F(u)=\int_0^u f(u')du'$; (2) momentum $P(u)=\frac{1}{2}\int u^2dx$ and (3) total mass $N(u)=\int udx$. Define the invariant

$$I(u)=E(u)+cP(u)-aN(u),$$

then $I'(u_{c,a})=0$ if and only if $u_{c,a}$ is a traveling wave solution satisfying (11.18). So

$$I(u)-I(u_{c,a})=\langle\mathcal{L}_0\delta u,\delta u\rangle+O\left(\|\delta u\|^3\right),\text{ where }\delta u=u-u_{c,a}. \tag{11.25}$$

Denote

$$X_1 = \left\{u \in H^{\frac{m}{2}} \mid \langle u, \mathcal{L}_0 U_{c,a}\rangle = \langle u, \mathcal{L}_0 \partial_c u_{c,a}\rangle = 0\right\}$$

to be the orthogonal complement of $X_2 = span\left\{U_{c,a}, \partial_c u_{c,a}\right\}$ in $\langle \mathcal{L}_0 \cdot, \cdot\rangle$. Since

$$\mathcal{L}_0 U_{c,a} = N'(u_{c,a}), \quad \mathcal{L}_0 \partial_c u_{c,a} = P'(u_{c,a}),$$

X_1 is the tangent space of the intersection of the level surfaces of the conserved momentum P and mass N. With D assumed to be nonsingular, X_2 roughly represents the gradient directions of P and N and thus $X = X_1 \oplus X_2$. Moreover, we have $\langle \mathcal{L}_0 \cdot, \cdot\rangle |_{X_1} \geq 0$ since

$$n^-\left(\mathcal{L}_0|_{X_1}\right) = n^-\left(\mathcal{L}_0\right) - n^-\left(\mathcal{L}_0|_{X_2}\right) = n^-\left(\mathcal{L}_0\right) - n^-\left(D\right) = 0.$$

We further decompose

$$X_1 = Y \oplus span\{\partial_x u_{c,a}\}, \quad \text{where } Y = \{u \in X_1 \mid (u, \partial_x u_{c,a})_X = 0\}.$$

Since $\ker \mathcal{L}_0 = span\{\partial_x u_{c,a}\}$, there exists $c_0 > 0$ such that $\langle \mathcal{L}_0 \delta u, \delta u\rangle \geq c_0 \left\|\delta u\right\|^2$ for any $\delta u \in Y$.

Suppose $u(t)$ is solution with $u(0)$ close to $u_{c,a}$ and $h(t) \in \mathbf{S}^1$ satisfies

$$\left\|u - u_{c,a}\left(\cdot - h\right)\right\| = \min_{y \in \mathbf{S}^1} \left\|u - u_{c,a}\left(\cdot - y\right)\right\|,$$

then $w(t) = u(t) - u_{c,a}\big(\cdot - h(t)\big) \in Y \oplus X_2$. By using the conservation of P and N to control the X_2 components of $w(t)$, and the uniform positivity of $\mathcal{L}_0$ on Y and (11.25) to control the Y component, we obtain the orbital stability. More details of such arguments can be found for example in [**29**] [**53**].

For BBM type equations, the Lyapunov functional is

$$I\left(u\right) = cP\left(u\right) - E\left(u\right) - caN\left(u\right),$$

where the energy functional $E\left(u\right) = \int \left(\frac{1}{2}u^2 + F\left(u\right)\right)\, dx$ and $P\left(u\right)$ is defined in (11.6). The rest of the proof is the same as in the KDV case. For good Boussinesq type equations (11.3), we write it as a first order Hamiltonian system

$$\partial_t \begin{pmatrix} u \\ v \end{pmatrix} = J\,\nabla E\left(u, v\right),$$

where $J = \begin{pmatrix} 0 & \partial_x \\ \partial_x & 0 \end{pmatrix}$ and the energy functional

$$E\left(u, v\right) = \frac{1}{2}\left(\mathcal{M}u, u\right) + \int \left(\frac{1}{2}v^2 + \frac{1}{2}u^2 - F\left(u\right)\right) dx.$$

For the traveling wave solution $\left(u_{c,a}\left(x - ct\right), v_{c,a}\left(x - ct\right)\right)$, $u_{c,a}$ satisfies (11.17) and $v_{c,a} = -cu_{c,a}$. Let $\vec{u} = \left(u, v\right)^T$ and construct the Lyapunov functional

$$I\left(\vec{u}\right) = E\left(\vec{u}\right) + cP\left(\vec{u}\right) - aN_1\left(\vec{u}\right),$$

where

$$P\left(\vec{u}\right) = \int uv\, dx, \; N_1\left(\vec{u}\right) = \int u dx, \; N_2\left(\vec{u}\right) = \int v dx.$$

Then $I'\left(\vec{u}_{c,a}\right) = 0$. The rest of the proof is the same. □

Compared with solitary waves, the periodic traveling waves have richer structures. They consist of a three parameter (period T, speed c, and integration constant a) family of solutions and different type of perturbations (co-periodic, multiple periodic, localized etc.) can be considered. In recent years, there have been lots of works on stability/instability of periodic traveling waves of dispersive PDEs. For co-periodic perturbations (i.e. of the same period), the nonlinear orbital stability were proved for various dispersive models (e.g. [**2**] [**3**] [**41**] [**34**] [**11**] [**8**]) by using Liapunov functionals. These stability results were proved for the cases when $\dim\ker(\mathcal{L}_0) = 1$ and $n^-(\mathcal{L}_0) = n^-(D)$ as in Proposition 11.1. An instability index formula similar to (11.24) was proved for KDV type equations ([**32**] [**43**] [**13**]). In these papers, some conditions (e.g. Assumption 2.1 in [**32**] and Assumption 3 in [**43**]) were imposed to ensure that the generalized eigenvectors of $J\mathcal{L}_0$ form a basis of X. These assumptions can be checked for the case $\mathcal{M} = -\partial_x^2$. In Theorem 11.2, we do not need such assumptions on the completion of generalized eigenspaces of $J\mathcal{L}_0$ and therefore we can get the index formula for very general nonlocal operators $\mathcal{M}$. In [**11**], an index formula was proved for periodic traveling waves of good Boussinesq equation ($\mathcal{M} = -\partial_x^2$) by using the theory of quadratic operator pencils. In [**8**], a parity instability criterion (as in Corollary 11.3 (ii)) was proved for periodic waves of several Hamiltonian PDEs including generalized KDV equations by using Evans functions.

Besides providing a unified way to get instability index formula and the stability criterion, we could also use the exponential trichotomy of $e^{J\mathcal{L}_0}$ in Theorem 11.2 to construct invariant manifolds near the orbit of unstable periodic traveling waves. Moreover, as in the case of solitary waves, when $\dim\ker(\mathcal{L}_0) = 1$, D is nonsingular and $k_i^- = 0$, we have orbital stability and local uniqueness of the center manifolds once constructed.

11.3. Modulational Instability of periodic traveling waves

Consider periodic traveling waves $u_{c,a}(x - ct)$ studied in Section 11.2. Assume the conditions in Proposition 11.1, so that $u_{c,a}$ is orbitally stable under perturbations of the same period. In this section, we consider modulational instability of periodic traveling waves, under perturbations of different period or even localized perturbations. The modulational instability, also called Benjamin-Feir or side-band instability in the literature, is a very important instability mechanism in lots of dispersive and fluid models. Again, we assume the minimal period of the traveling wave $u_{c,a}$ is 2π. We focus on KDV type equations (11.2), and the consideration for BBM and good-Boussinesq type equations is similar. We assume the Fourier symbol $\alpha(\xi)$ of the operator $\mathcal{M}$ is even, so that $\mathcal{M}$ is a real operator. Based on the standard Floquet-Bloch theory, we seek bounded eigenfunction $\phi(x)$ of the linearized operator $J\mathcal{L}_0$ in the form of $\phi(x) = e^{ikx}v_k(x)$, where $k \in \mathbf{R}$ is a parameter and $v_k \in L^2(\mathbf{S}^1)$. Recall that $J = \partial_x$ and $\mathcal{L}_0 := \mathcal{M} + c - f'(u_{c,a})$. It leads us to the one-parameter family of eigenvalue problems

$$J\mathcal{L}_0 e^{ikx}v_k(x) = \lambda(k)e^{ikx}v_k(x),$$

or equivalently $\mathcal{J}_k\mathcal{L}_k v_k = \lambda(k)\, v_k$, where

$$\mathcal{J}_k = \partial_x + ik,\ \mathcal{L}_k = \mathcal{M}_k + c - f'(u_{c,a}). \tag{11.26}$$

Here, $\mathcal{M}_k$ is the Fourier multiplier operator with the symbol $\alpha(\xi+k)$. We say that $u_{c,a}$ is linearly modulationally unstable if there exists $k \notin \mathbf{Z}$ such that the operator $\mathcal{J}_k\mathcal{L}_k$ has an unstable eigenvalue $\lambda(k)$ with $\operatorname{Re}\lambda(k) > 0$ in the space $L^2(\mathbf{S}^1)$.

Since $\mathcal{J}_k$ and $\mathcal{L}_k$ are complex operators, we first reformulate the problem in terms of real operators to use the general theory in this paper. Consider

$$\phi(x) = \cos(kx)\, u_1(x) + \sin(kx)\, u_2(x), \tag{11.27}$$

where $u_1, u_2 \in L^2(\mathbf{S}^1)$ are real functions. By definition,

$$\mathcal{M}\left(e^{ikx}u(x)\right) = e^{ikx}\mathcal{M}_k u.$$

We decompose

$$\mathcal{M}_k = \mathcal{M}_k^e + i\mathcal{M}_k^o, \quad \mathcal{M}_{-k} = \mathcal{M}_k^e - i\mathcal{M}_k^o$$

where $\mathcal{M}_k^e, \mathcal{M}_k^o$ are operators with Fourier multipliers

$$\alpha_k^e(\xi) = \frac{1}{2}\left(\alpha(\xi+k) + \alpha(\xi-k)\right)$$

and

$$\alpha_k^o(\xi) = -\frac{i}{2}\left(\alpha(\xi+k) - \alpha(\xi-k)\right).$$

Then $\mathcal{M}_k^e$, $\mathcal{M}_k^o$ are self-adjoint and skew-adjoint respectively. Since $\overline{\alpha_k^{e,o}(\xi)} = \alpha_k^{e,o}(-\xi)$, $\mathcal{M}_k^e$ and $\mathcal{M}_k^o$ map real functions to real. In particular, for $\mathcal{M} = -\partial_x^2$, we have $\mathcal{M}_k^e = -\partial_x^2 + k^2$ and $\mathcal{M}_k^o = -2k\partial_x$. By using

$$\phi(x) = \frac{e^{ikx}}{2}(u_1 - iu_2) + \frac{e^{-ikx}}{2}(u_1 + iu_2) \tag{11.28}$$

and via simple computations, we obtain

$$\mathcal{M}\phi = \cos(kx)\left(\mathcal{M}_k^e u_1 + \mathcal{M}_k^o u_2\right) + \sin(kx)\left(-\mathcal{M}_k^o u_1 + \mathcal{M}_k^e u_2\right),$$

and

$$J\phi = \cos(kx)\left(\partial_x u_1 + ku_2\right) + \sin(kx)\left(\partial_x u_2 - ku_1\right).$$

Define the operators

$$J_k = \begin{pmatrix} \partial_x & k \\ -k & \partial_x \end{pmatrix}, \quad L_k = \begin{pmatrix} \mathcal{M}_k^e + c - f'(u_{c,a}) & \mathcal{M}_k^o \\ -\mathcal{M}_k^o & \mathcal{M}_k^e + c - f'(u_{c,a}) \end{pmatrix}. \tag{11.29}$$

Then J_k, L_k are skew-adjoint and self-adjoint real operators and

$$\mathcal{J}\mathcal{L}_0\phi = \left(\cos(kx), \sin(kx)\right) J_k L_k \begin{pmatrix} u_1 \\ u_2 \end{pmatrix}.$$

As always in the spectral analysis, u_1 and u_2, as well as operator J_kL_k and quadratic forms $\langle L_k\cdot,\cdot\rangle$ and $\langle\cdot, J_k\cdot\rangle$, need to be complexified. By using operators $\mathcal{J}_k$ and $\mathcal{L}_k$ we can diagonalize J_kL_k and L_k blockwisely. In fact, let

$$w_1 = \frac{1}{2}(u_1 - iu_2), \quad w_2 = \frac{1}{2}(u_1 + iu_2), \quad S\begin{pmatrix} u_1 \\ u_2 \end{pmatrix} = \begin{pmatrix} w_1 \\ w_2 \end{pmatrix}.$$

One may compute using (11.28) and the definition of $\mathcal{J}_k$ and $\mathcal{L}_k$

$$L_k = S^{-1}\begin{pmatrix} \mathcal{L}_k & 0 \\ 0 & \mathcal{L}_{-k} \end{pmatrix} S, \quad J_kL_k = S^{-1}\begin{pmatrix} \mathcal{J}_k\mathcal{L}_k & 0 \\ 0 & \mathcal{J}_{-k}\mathcal{L}_{-k} \end{pmatrix} S. \tag{11.30}$$

Moreover, $\mathcal{L}_{-k}$ and $\mathcal{J}_{-k}\mathcal{L}_{-k}$ are the complex conjugates of L_k and J_kL_k respectively, namely,

$$\mathcal{L}_{-k}w = \overline{\mathcal{L}_k\bar{w}}, \quad \mathcal{J}_{-k}\mathcal{L}_{-k}w = \overline{\mathcal{J}_k\mathcal{L}_k\bar{w}}. \tag{11.31}$$

From the above relations we obtain

$$n^-(L_k) = n^-(\mathcal{L}_k) + n^-(\mathcal{L}_{-k}) = 2n^-(\mathcal{L}_k),$$

where $n^-(\mathcal{L}_k)$ is understood as the negative index of the complex Hermitian form $\langle \mathcal{L}_k \cdot, \cdot \rangle$. Moreover, (11.31) implies that $\lambda \in \sigma(\mathcal{J}_k \mathcal{L}_k)$ if and only if $\bar{\lambda} \in \sigma(\mathcal{J}_{-k}\mathcal{L}_{-k})$, with $\ker(\bar{\lambda} - \mathcal{J}_{-k}\mathcal{L}_{-k})^n$ consisting exactly of the complex conjugates of the functions in $\ker(\lambda - \mathcal{J}_k \mathcal{L}_k)^n$ for any $n > 0$. Next, it is easy to see that $J_k L_k$, as well as $\mathcal{J}_k \mathcal{L}_k$, has compact resolvents and thus $\sigma(J_k L_k)$, as well as $\sigma(\mathcal{J}_k \mathcal{L}_k)$, consists of only discrete eigenvalues of finite algebraic multiplicity. Therefore, by Proposition 2.3, for any purely imaginary eigenvalue $i\mu \in \sigma(J_k L_k)$, L_k is non-degenerate on the finite dimensional eigenspace $E_{i\mu}$, and thus $n^{\leq 0}(L_k|_{E_{i\mu}}) = n^-(L_k|_{E_{i\mu}})$. Let (k_r, k_c, k_i^-, k_0^-) be the indices defined in (2.10), (2.11), and (2.12) for $J_k L_k$, and $(\tilde{k}_r, \tilde{k}_0^-)$ be the corresponding indices for the positive and zero eigenvalues of $\mathcal{J}_k \mathcal{L}_k$. Let $\tilde{k}_c$ be the sum of algebraic multiplicities of eigenvalues of $\mathcal{J}_k \mathcal{L}_k$ in the first and the fourth quadrants, $\tilde{k}_i^-$ be the total number of negative dimensions of $\langle \mathcal{L}_k \cdot, \cdot \rangle$ restricted to the subspaces of generalized eigenvectors of nonzero purely imaginary eigenvalues of $\mathcal{J}_k \mathcal{L}_k$. On the one hand, (11.30) and (11.31) imply

$$k_r = 2\tilde{k}_r, \quad k_c = \tilde{k}_c, \quad k_i^- = \tilde{k}_i^-, \quad k_0^- = 2\tilde{k}_0^-.$$

On the other hand, Theorem 2.3 implies

$$k_r + 2k_c + 2k_i^- + k_0^- = 2n^-(\mathcal{L}_k)\,. \tag{11.32}$$

Therefore we obtain

PROPOSITION 11.2. *For any $k \in (0,1)$,*

$$\tilde{k}_r + \tilde{k}_c + \tilde{k}_i^- + \tilde{k}_0^- = n^-(\mathcal{L}_k)\,. \tag{11.33}$$

The modulational instability occurs if $\tilde{k}_r \neq 0$ or $\tilde{k}_c \neq 0$.

REMARK 11.1. *Note that $\mathcal{J}_k$ is invertible for any $k \notin \mathbf{Z}$. With a more concrete form of $\mathcal{M}$, it is possible to determine $\tilde{k}_0^-$.*

- *Firstly, if $\ker \mathcal{L}_0$ is known (recall $\partial_x u_{c,a} \in \ker \mathcal{L}_0$), then one may study $\ker \mathcal{L}_k$, as well as $\tilde{k}_0^-$, for $0 < |k| << 1$ through asymptotic analysis.*
- *If $\mathcal{M} = -\partial_{xx}$, then $\ker \mathcal{L}_k = \{0\}$ for any $k \in (0,1)$ (and thus for any $k \notin \mathbf{Z}$). In fact, in this case,*

$$\mathcal{L}_0 = -\partial_{xx} + c - f'(u_{c,a}), \quad v \in \ker \mathcal{L}_k \iff e^{ikx}v \in \ker \mathcal{L}_0$$

and $\ker \mathcal{L}_0 = span\{\partial_x u_{c,a}\}$. Suppose $\mathcal{L}_k$ has nontrivial kernel for some $k \in (0,1)$ and $0 \neq v \in \ker \mathcal{L}_k$. Denote $v_0 \triangleq \partial_x u_{c,a}$, then the Wronskian of v_0 and $e^{ikx}v$ satisfies

$$W(x) = e^{ikx}(v_x v_0 - v v_{0x} + ikvv_0) = const.$$

Since v and v_0 are 2π-periodic and $k \in (0,1)$, it must hold that

$$v_x v_0 - v v_{0x} + ikvv_0 = 0. \tag{11.34}$$

We claim $v(x) \neq 0$ for any $x \in \mathbf{S}^1$. In fact, if $v(x_0) = 0$, then $v_x(x_0) \neq 0$ and (11.34) imply $v_0(x_0) = 0$. The uniqueness of the solution to the ODE $\mathcal{L}_0 u = 0$ leads to the proportionality between v_0 and $e^{ikx}v$, a contradiction to $k \in (0,1)$ and the 2π-periodicity of $v(x)$. Now that $v(x) \neq 0$, (11.34) implies $\frac{v_0}{v} = Ce^{-ikx}$, which is again a contradiction.

REMARK 11.2. *The above index formula (11.33) was proved in* [**32**] *for the case when* $\ker \mathcal{L}_k = \{0\}$, *with additional assumptions to ensure that the generalized eigenfunctions of* $\mathcal{J}_k\mathcal{L}_k$ *form a complete basis of* $L^2(\mathbf{S}^1)$ *as assumed in the case of co-periodic perturbations. Proposition 11.2 is proved without such assumptions.*

REMARK 11.3. *We can also consider the case when the operator* $\mathcal{M}$ *is a smoothing operator, that is,* $\|\mathcal{M}(\cdot)\|_{H^r} \sim \|\cdot\|_{L^2}$ *for some* $r > 0$. *One example is the Whitham equation which is a KDV type equation (11.2) with the symbol of* $\mathcal{M}$ *being* $\sqrt{\frac{\tanh \xi}{\xi}}$ *and thus* $r = \frac{1}{2}$. *In this case, if we assume that*

$$-c - \|f'(u_c)\|_{L^\infty(\mathbb{T}_{2\pi})} \geqslant \epsilon > 0, \tag{11.35}$$

then $\mathcal{L}_0$ *and* $\mathcal{L}_k$ *are compact perturbations of the positive operator* $-c + f'(u_{c,a})$ *so that* $n^-(-\mathcal{L}_0), n^-(-\mathcal{L}_k) < \infty$. *Then the index formula*

$$\bar{k}_r + \bar{k}_c + k_i^- + k_0^- = n^-(-\mathcal{L}_k)$$

is still true for the operator $\mathcal{J}_k\mathcal{L}_k$, $k \in (0,1)$. *The assumption (11.35) can be verified (*[**36**]*) for small amplitude periodic traveling waves of Whitham equation with* $f(u) = u^2$.

Under the conditions of orbital stability in Proposition 11.1, the spectrum of the operator $J\mathcal{L}_0$ in $L^2(\mathbf{S}^1)$ lie on the imaginary axis and are all discrete. Moreover, the non-degeneracy of the matrix D (defined by (11.21)) implies that the generalized kernel of $J\mathcal{L}_0$ is spanned by $\{\partial_x u_{c,a}, \partial_c u_{c,a}, U_{c,a}\}$. For $k \in (0,1)$ small, it is natural to study the spectra of $\mathcal{J}_k\mathcal{L}_k$ by the perturbation theory. Even though the results in Section 2.5 and Chapter 9 do not apply directly as $\mathcal{J}_k - J : X^* \to X$ is not bounded, the ideas there and the property that $\mathcal{J}_k\mathcal{L}_k$ has only isolated eigenvalues still yield the desired results. We start with the following lemma on the resolvent of $\mathcal{J}_k\mathcal{L}_k$.

LEMMA 11.1. *Assume that the symbol* $\alpha(\xi)$ *of* $\mathcal{M}$ *satisfies* $a|\xi|^m \le \alpha(\xi) \le b|\xi|^m$, $a, b > 0$, $m > 0$, *for large* ξ *and*

$$\limsup_{\rho\to 0}\sup_{\xi\in\mathbf{Z}} \frac{|\alpha(\xi+\rho) - \alpha(\xi)|}{1+|\xi|^m} \to 0, \tag{11.36}$$

then the resolvent $(\lambda - \mathcal{J}_k\mathcal{L}_k)^{-1}$ *is continuous in* $k \in [0,1]$.

PROOF. Fix $k \in [0,1]$. From (11.26), one can compute

$$\mathcal{J}_{k'}\mathcal{L}_{k'} - \mathcal{J}_k\mathcal{L}_k = (\partial_x + ik)(\mathcal{M}_{k'} - \mathcal{M}_k) + i(k'-k)\big(\mathcal{M}_{k'} + c - f'(u_{c,a})\big).$$

On the one hand, there exists $a_0 \neq 0$ such that $a_0 + (\partial_x + ik)\mathcal{M}_k$ has a compact inverse on X. We obtain from (11.36)

$$|\big(a_0 + (\partial_x + ik)\mathcal{M}_k\big)^{-1}(\mathcal{J}_{k'}\mathcal{L}_{k'} - \mathcal{J}_k\mathcal{L}_k)| \to 0 \ \text{ as } k' \to k. \tag{11.37}$$

On the other hand, (11.26) and $m > 0$ imply that

$$\begin{aligned} &I + \big(a_0 + (\partial_x + ik)\mathcal{M}_k\big)^{-1}(\lambda - \mathcal{J}_k\mathcal{L}_k) \\ =&\big(I + (\partial_x + ik)\mathcal{M}_k\big)^{-1}(\lambda + a_0 - (\partial_x + ik)(c - f'(u_{c,a})) \end{aligned}$$

is compact. Therefore $A = \big(a_0 + (\partial_x + ik)\mathcal{M}_k\big)^{-1}(\lambda - \mathcal{J}_k\mathcal{L}_k)$ is a Fredholm operator of index 0. Suppose $\lambda \notin \sigma(\mathcal{J}_k\mathcal{L}_k)$, then A is injective and thus A^{-1} is bounded on

X. Along with (11.37), we obtain

$$|(\lambda - \mathcal{J}_k\mathcal{L}_k)^{-1}(\mathcal{J}_{k'}\mathcal{L}_{k'} - \mathcal{J}_k\mathcal{L}_k)| = |A^{-1}\big(a_0 + (\partial_x + ik)\mathcal{M}_k\big)^{-1}(\mathcal{J}_{k'}\mathcal{L}_{k'} - \mathcal{J}_k\mathcal{L}_k)| \to 0$$

as $k' \to k$. From

$$\lambda - \mathcal{J}_{k'}\mathcal{L}_{k'} = (\lambda - \mathcal{J}_k\mathcal{L}_k)\big(I - (\lambda - \mathcal{J}_k\mathcal{L}_k)^{-1}(\mathcal{J}_{k'}\mathcal{L}_{k'} - \mathcal{J}_k\mathcal{L}_k)\big),$$

we obtain the continuity of the resolvent $(\lambda - \mathcal{J}_k\mathcal{L}_k)^{-1}$ in $k \in [0,1]$. □

REMARK 11.4. *The assumption* (11.36) *is clearly satisfied if* $\alpha(\xi) \in C^1(\mathbf{R})$ *and*

$$\limsup_{|\xi|\to\infty} \frac{\alpha'(\xi)}{|\xi|^m} < \infty.$$

Next we show that when k is small enough, the unstable modes of $\mathcal{J}_k\mathcal{L}_k$ can only bifurcate from the zero eigenvalue of $J\mathcal{L}_0$.

PROPOSITION 11.3. *Suppose* $\ker\mathcal{L}_0 = span\{\partial_x u_{c,a}\}$, D *is nonsingular,* $n^-(\mathcal{L}_0) = n^-(D)$ *and* (11.36) *holds. Then for any* $\delta > 0$, *there exists* $\varepsilon_0 > 0$ *such that if* $|k| < \varepsilon_0$, *then* $\sigma(\mathcal{J}_k\mathcal{L}_k) \cap \{|z| \geq \delta\} \subset i\mathbf{R}$.

PROOF. Since 0 is an isolated spectral point of $J\mathcal{L}_0$, there exists $\delta_0 > 0$ such that $\lambda \notin \sigma(J\mathcal{L}_0)$ as long as $0 < |\lambda| \leq \delta_0$. Without loss of generality, assume $0 < \delta < \delta_0$. Lemma 11.1 implies $\lambda \notin \sigma(\mathcal{J}_k\mathcal{L}_k)$ for $0 < |k| << 1$. Let

$$P(k) = \frac{1}{2\pi i}\oint_{|\lambda|=\delta}(\lambda - \mathcal{J}_k\mathcal{L}_k)^{-1}d\lambda, \quad Z_k = P(k)X, \quad Y_k = \big(I - P(k)\big)X.$$

The standard spectral theory implies that $P(k)$ is continuous in k, Y_k and Z_k are invariant under $\mathcal{J}_k\mathcal{L}_k$, and

$$|\lambda| < \delta,\ \forall\lambda \in \sigma(\mathcal{J}_k\mathcal{L}_k|_{Z_k}) \text{ and } |\lambda| > \delta,\ \forall\lambda \in \sigma(\mathcal{J}_k\mathcal{L}_k|_{Y_k}).$$

For $k = 0$, our assumptions imply that $Z_0 = span\{\partial_x u_{c,a}, \partial_c u_{c,a}, U_{c,a}\}$. Therefore Z_k close to Z_0 is a 3-dim invariant subspace of $\mathcal{J}_k\mathcal{L}_k$ with small eigenvalues containing $\ker\mathcal{L}_k$. Moreover, the assumption

$$n^-(\mathcal{L}_0) = n^-(D) = n^-(\mathcal{L}_0|_{Z_0})$$

and the $\mathcal{L}_0$-orthogonality between Z_0 and Y_0 imply that $\mathcal{L}_0$ is uniformly positive definite on Y_0. As $\mathcal{L}_k : X = H^{\frac{m}{2}} \to X^* = H^{-\frac{m}{2}}$ is continuous in k, there exists $\alpha > 0$ such that $\langle\mathcal{L}_k u, u\rangle > \alpha\|u\|^2$ for all $u \in Y_k$. Clearly $\mathcal{J}_k\mathcal{L}_k|_{Y_k}$ is skew-adjoint with respect to the equivalent inner product given by $\langle\mathcal{L}_k\cdot,\cdot\rangle$ on Y_k, therefore $\sigma(\mathcal{J}_k\mathcal{L}_k|_{Y_k}) \subset i\mathbf{R}$ and the proposition follows. □

Since $\dim\ker(J\mathcal{L}_0) = 3$, the perturbation of zero eigenvalue of $J\mathcal{L}_0$ for $\mathcal{J}_k\mathcal{L}_k$ $(0 < k \ll 1)$ can be reduced to the eigenvalue perturbation of a 3 by 3 matrix. This had been studied extensively in the literature and instability conditions were obtained for various dispersive models. See the survey [**12**] and the references therein.

Recently, it was proved in [**36**] that linear modulational instability of the traveling wave $u_c(x - ct)$ also implies the nonlinear instability for both multi-periodic and localized perturbations. The semigroup estimates of $e^{tJ\mathcal{L}_0}$ play an important role on this proof of nonlinear instability. We sketch these estimates below, as an example of the application of Theorem 2.2 on the exponential trichotomy of linear Hamiltonian PDE. First, if u_c is linearly modulationally unstable, then there exists a rational $k_0 = \frac{p}{q} \in (0,1)$ such that $\mathcal{J}_{k_0}\mathcal{L}_{k_0}$ has an unstable eigenvalue. By the definition of $\mathcal{J}_{k_0}\mathcal{L}_{k_0}$, this implies that the operator $J\mathcal{L}_0$ has an unstable eigenvalue on the $2\pi q$ periodic space $L^2\left(\mathbf{S}^1_{2\pi q}\right)$ with an eigenfunction of the form $e^{ik_0x}u\left(x\right)$ $\left(u \in L^2\left(\mathbf{S}^1\right)\right)$. The exponential trichotomy of the semigroup $e^{tJ\mathcal{L}_0}$ on the space $H^s\left(\mathbf{S}^1_{2\pi q}\right)$ $\left(s \geq \frac{m}{2}\right)$ follows directly by Theorem 2.2. This is used in [**36**] to prove nonlinear orbital instability of u_c for $2\pi q$ periodic perturbations or even to construct stable and unstable manifolds. To prove nonlinear instability for localized perturbations, we study the semigroup $e^{tJ\mathcal{L}_0}$ on the space $H^s\left(\mathbf{R}\right)$ $\left(s \geq \frac{m}{2}\right)$. The operator $\mathcal{L}_0$ might have negative continuous spectrum in $H^s\left(\mathbf{R}\right)$. For example, when $\mathcal{M} = -\partial_x^2$, the spectrum of $\mathcal{L}_0 = -\partial_x^2 + V\left(x\right)$ with periodic $V\left(x\right)$ is well studied in the literature and is known to have bands of continuous spectrum. So Theorem 2.2 does not apply. However, we have the following upper bound estimate of $e^{tJ\mathcal{L}_0}$ on $H^s\left(\mathbf{R}\right)$, which suffices to prove nonlinear localized instability.

LEMMA 11.2. *Assume* (11.36). *Let* $\lambda_0 \geq 0$ *be such that*

$$\operatorname{Re}\lambda \leq \lambda_0, \quad \forall \xi \in [0,1],\ \lambda \in \sigma(\mathcal{J}_\xi\mathcal{L}_\xi).$$

For every $s \geq \frac{m}{2}$, *there exist* $C(s) > 0$ *such that*

$$\|e^{t\mathcal{J}_\xi\mathcal{L}_\xi}v(x)\|_{H^s(\mathbf{S}^1)} \leq C(s)(1+t^{2n^-(\mathcal{L}_\xi)+1})e^{\lambda_0 t}\|v(x)\|_{H^s(\mathbf{S}^1)}, \tag{11.38}$$

$$\|e^{tJ\mathcal{L}_0}u(x)\|_{H^s(\mathbf{R})} \leqslant C(s)(1+t^{2n^-(\mathcal{L}_\xi)+1})e^{\lambda_0 t}\|u(x)\|_{H^s(\mathbf{R})}, \tag{11.39}$$

for any $\xi \in [0,1]$, $v \in H^s(\mathcal{S}^1)$, *and* $u \in H^s(\mathbf{R})$.

PROOF. It suffices to prove the lemma for $s = \frac{m}{2}$. The estimates for general $s \geq \frac{m}{2}$ can be obtained by applying $\mathcal{J}_\xi\mathcal{L}_\xi$ and $J\mathcal{L}_0$ repeatedly to the estimates for $s = \frac{m}{2}$ (and interpolation for the case when $\frac{2s}{m}$ is not an integer). We start with the first estimate in the 2π-periodic case. Due to the compactness of $[0,1]$, it suffices to prove that for any $\xi_0 \in [0,1]$, there exist $C, \epsilon > 0$ and an integer $K \geq 0$ such that (11.38) holds for $\xi \in (\xi_0 - \epsilon, \xi_0 + \epsilon)$. We first note that each $\lambda \in \sigma(\mathcal{J}_{\xi_0}\mathcal{L}_{\xi_0})$ is an isolated eigenvalue with finite algebraic multiplicity and $\mathcal{L}_{\xi_0}$ is non-degenerate on $E_\lambda/(E_\lambda \cap \ker \mathcal{L}_{\xi_0})$. Let

$$\Lambda = \{\lambda \in \sigma(\mathcal{J}_{\xi_0}\mathcal{L}_{\xi_0}) \mid \exists\, \delta > 0 \text{ s.t. } \langle \mathcal{L}_{\xi_0}v, v\rangle \geq \delta\|v\|^2\}.$$

Due to Proposition 11.2, $\sigma(\mathcal{J}_{\xi_0}\mathcal{L}_{\xi_0})\backslash\Lambda$ is finite and

$$n = \Sigma_{\lambda\in\sigma(\mathcal{J}_{\xi_0}\mathcal{L}_{\xi_0})\backslash\Lambda} \dim E_\lambda < \infty.$$

Moreover, there exists $\varepsilon > 0$ such that

$$\Omega \cap \Lambda = \emptyset, \quad \text{where } \Omega = \cup_{\lambda\in\sigma(\mathcal{J}_{\xi_0}\mathcal{L}_{\xi_0})\backslash\Lambda}\{z \mid \|z - \lambda\| < \varepsilon\} \subset \mathbf{C}.$$

From Lemma 11.1, there exists $\epsilon > 0$ such that $\partial\Omega \cap \sigma(\mathcal{J}_\xi\mathcal{L}_\xi) = \emptyset$ for any $\xi \in [\xi_0 - \epsilon, \xi_0 + \epsilon]$. For such ξ, let

$$P(\xi) = \frac{1}{2\pi i}\oint_{\partial\Omega}(\lambda - \mathcal{J}_k\mathcal{L}_k)^{-1}d\lambda, \quad Z_\xi = P(\xi)X, \quad Y_\xi = (I - P(\xi))X,$$

which are continuous in ξ and invariant under $e^{t\mathcal{J}_\xi\mathcal{L}_\xi}$. Therefore $\dim Z_\xi = n$ and the continuity of $\mathcal{L}_\xi$ in ξ implies that there exists $\delta > 0$ such that

$$\delta^{-2}\|v\|^2 \geq \langle \mathcal{L}_\xi v, v\rangle \geq \delta^2\|v\|^2, \quad \forall v \in Y_\xi,\ |\xi - \xi_0| \leq \epsilon.$$

Moreover, according to Proposition 2.2, for any $\lambda \in \Omega \cap \sigma(\mathcal{J}_\xi \mathcal{L}_\xi)$, the dimension of its eigenspace

$$E_\lambda(\mathcal{J}_\xi \mathcal{L}_\xi) = \ker(\lambda - \mathcal{J}_\xi \mathcal{L}_\xi)^{2\left(1+n^-(\mathcal{L}_\xi)\right)},$$

namely, the maximal dimension of Jordan blocks of $\mathcal{J}_\xi \mathcal{L}_\xi$ on Y_ξ is no more than $2(1+n^-(\mathcal{L}_\xi))$. So for any $\xi \in [\xi_0 - \epsilon, \xi_0 + \epsilon]$, there exists a generic constant $C > 0$ independent of ξ, such that

$$\begin{aligned}
&\|e^{t\mathcal{J}_\xi \mathcal{L}_\xi} v\| \leq \|e^{t\mathcal{J}_\xi \mathcal{L}_\xi} P(\xi) v\| + \|e^{t\mathcal{J}_\xi \mathcal{L}_\xi}(I - P(\xi))v\| \\
\leq &C\Big((1+t^{2n^-(\mathcal{L}_\xi)+1})e^{\lambda_0 t}\|P(\xi)v\| + \langle \mathcal{L}_\xi e^{t\mathcal{J}_\xi \mathcal{L}_\xi}(I-P(\xi))v, e^{t\mathcal{J}_\xi \mathcal{L}_\xi}(I-P(\xi))v\rangle^{\frac{1}{2}}\Big) \\
\leq &C\Big((1+t^{2n^-(\mathcal{L}_\xi)+1})e^{\lambda_0 t}\|P(\xi)v\| + \langle \mathcal{L}_\xi (I-P(\xi))v, (I-P(\xi))v\rangle^{\frac{1}{2}}\Big) \\
\leq &C(1+t^{2n^-(\mathcal{L}_\xi)+1})e^{\lambda_0 t}\|v\|.
\end{aligned}$$

Along with the compactness of $[0,1]$, it implies (11.38).

To prove (11.39), we first write, for any $u \in H^s(\mathbf{R})$,

$$u(x) = \int_0^1 e^{i\xi x} u_\xi(x) d\xi, \quad \text{where } u_\xi(x) = \Sigma_{n\in\mathbf{Z}} e^{inx}\hat{u}(n+\xi) \in H^s(\mathbf{S}^1),$$

and $\hat{u}$ is the Fourier transform of u. Clearly there exists $C > 0$ such that

$$\frac{1}{C}\|u\|^2_{H^s(\mathbb{R})} \leq \int_0^1 \|u_\xi(x)\|^2_{H^s(\mathbf{S}^1)} d\xi \leq C\|u\|^2_{H^s(\mathbb{R})}. \tag{11.40}$$

Note

$$e^{tJ\mathcal{L}_0}u(x) = \int_0^1 e^{i\xi x} e^{t\mathcal{J}_\xi \mathcal{L}_\xi} u_\xi(x)\, d\xi$$

and thus

$$\|e^{tJ\mathcal{L}_0}u(x)\|^2_{H^s(\mathbf{R})} \approx \int_0^1 \|e^{t\mathcal{J}_\xi \mathcal{L}_\xi} u_\xi(x)\|^2_{H^s(\mathbf{S}^1)} d\xi. \tag{11.41}$$

Along with (11.38), it immediately implies (11.39). □

REMARK 11.5. *The semigroup estimates of the types (11.38) and (11.39) can also be obtained for $s = -1$, that is, in the negative Sobolev space $H^{-1}\left(\mathbf{S}^1_{2\pi q}\right)$ and $H^{-1}(\mathbf{R})$ for $e^{tJ\mathcal{L}_0}$ (see* [**36**]*). Such semigroup estimates were used in* [**36**] *to prove nonlinear modulational instability by a bootstrap argument.*

11.4. The spectral problem $Lu = \lambda u'$

In this section, we consider the eigenvalue problem of the form

$$Lu = \lambda u', \tag{11.42}$$

where the symmetric operator L is of the form of $\mathcal{L}_0$ in Section 11.1. As an example, consider the stability of solitary waves of generalized Bullough–Dodd equation ([**69**])

$$u_{tx} = au - f(u), \tag{11.43}$$

where $a > 0$ and f is a smooth function of u satisfying

$$f(u) = O\left(u^2\right),\ f'(u) = O(u) \text{ for small } u. \tag{11.44}$$

The traveling wave $u_c(x+ct)$ satisfies the ODE

$$-cu_c'' + au_c - f(u_c) = 0.$$

Then the linearized equation in the traveling frame $(x+ct,t)$ takes the form

$$u_{tx} = -cu_{xx} + au - f'(u_c)\,u. \tag{11.45}$$

Thus the eigenvalue problem takes the form (11.42) with

$$L = -c\frac{d^2}{dx^2} + a - f'(u_c)\,. \tag{11.46}$$

We consider the general problem (11.42) with L of the form $L = \mathcal{M}+V(x)$. We assume that: i) M is a Fourier multiplier operator with the symbol $\alpha(\xi)$ satisfying

$$\alpha(\xi) \geq 0 \text{ and } \alpha(\xi) \approx |\xi|^{2s} \ (s>0),\text{ when } |\xi| \text{ is large,} \tag{11.47}$$

and ii) the real potential $V(x)$ satisfies

$$V(x) \to \delta_0 > 0 \text{ when } |x| \to \infty. \tag{11.48}$$

Let $X = H^s(R)$ $(s>0)$. Then the assumption (**H2**) is satisfied for L on X. Namely, $L : X \to X^*$ is bounded and symmetric, and there exists a decomposition of X

$$X = X_- \oplus \ker L \oplus X_+, \quad n^-(L) \triangleq \dim X_- < \infty,$$

satisfying $L|_{X_-} < 0$ and $L|_{X_+} \geq \delta > 0$.

Define $J = \partial_x^{-1}$. Now we check that $J : X^* \to X$ is densely defined and $J^* = -J$. On $X = H^s(R)$ with $s>0$, the operator $\partial_x : X \to X^*$ is densely defined and satisfies $(\partial_x)^* = -\partial_x$. Since $\ker \partial_x = \{0\}$,

$$\overline{R(\partial_x)} = (\ker(\partial_x^*))^{\perp} = (\ker(-\partial_x))^{\perp} = X^*,$$

so $D(\partial_x^{-1}) = R(\partial_x)$ is dense in X^* and $J = \partial_x^{-1} : X^* \to X$ satisfies $J^* = -J$.

So the eigenvalue problem $Lu = \lambda u'$ can be equivalently written in the Hamiltonian form $JLu = \lambda u$, where (J,L,X) satisfies the assumptions (**H1**)-(**H3**). Let $\ker L = span\{\psi_1, \cdots, \psi_l\}$ and

$$span\{\psi_1', \cdots, \psi_l'\} \cap R(L) = span\{g_1, \cdots, g_m\}\,,\ m \leq l.$$

Define the m by m matrix

$$D = \left(\left\langle L^{-1}g_i, g_j\right\rangle\right),\ 1 \leq i,j \leq m.$$

By Theorem 2.3 and Proposition 2.7, we get the following theorem.

THEOREM 11.3. *Assume (11.47) and (11.48). Then*

$$k_r + 2k_c + 2k_i^{\leq 0} + k_0^{\leq 0} = n^-(L)\,,$$

where $k_r, k_c, k_i^{\leq 0}, k_0^{\leq 0}$ are the indexes for the eigenvalues of $\partial_x^{-1}L$, as defined in Section 2.4. In addition, we have $k_0^{\leq 0} \geq n^{\leq 0}(D)$, where $n^{\leq 0}(D)$ is the number of nonpositive eigenvalues of D. If D is nonsingular, then $k_0^{\leq 0} = n^-(D)$, i.e., the number of negative eigenvalues of D.

For many applications, particularly the generalized Bullough–Dodd equation where $\mathcal{M} = -c\partial_x^2$ $(c>0)$, L has at most one dimensional kernel and negative eigenspace. In this case, we get a more explicit instability criterion.

COROLLARY 11.4. *i) Assume $n^-(L) = 1$ and $\ker L = \{\psi_0\}$. Then there is a positive eigenvalue of $\partial_x^{-1}L$ when $\left\langle L^{-1}\psi_0', \psi_0'\right\rangle > 0$.*

ii) Assume $n^-(L) \leq 1$ and there exists $0 \neq \psi_0 \in \ker L$ such that $\left\langle L^{-1}\psi_0', \psi_0'\right\rangle \leq 0$, then $\partial_x^{-1}L$ has no unstable eigenvalues.

REMARK 11.6. *The above Corollary was obtained in* [**69**] *under some additional assumptions. In* [**69**], *Corollary* 11.4 *i) was proved under the following two assumptions:*

C1) $(f_0, g_0) \neq 0$, *where* f_0 *is the eigenfunction of* L *with the negative eigenvalue and* $g_0' \in \ker L$.

C2) For any $\lambda \in \mathbf{R}$,

$$\left\| P_+ (L - \lambda \partial_x)^{-1} P_+ v \right\|_{H^1} \leq C(\lambda) \|v\|_{L^2},$$

where P_+ *is the projection to the positive space of* L *and* $C(\lambda)$ *is bounded on compact sets.*

The proof in [**69**] *is by constructing Evans-like functions. Corollary* 11.4 *ii) was proved in* [**69**] *under the following additional assumptions:*

D1) $\ker L = \{\psi_0\}$ *and* $\langle L^{-1}\psi_0', \psi_0' \rangle < 0$;

D2) For any $\lambda \notin i\mathbf{R}$, *the operator* $L - \lambda\partial_x$ *has zero index and the equation* $(L - \lambda\partial_x) f = g$ *satisfies certain Fredholm alternative properties (see (12)(13)(14) in* [**69**]);

D3) The symbol $\alpha(\xi)$ *of the leading order part* $\mathcal{M}$ *of* L *satisfies*

$$\alpha(\xi) \approx |\xi|^{2s} \left(s > \frac{1}{2}\right), \text{ when } |\xi| \text{ is large.}$$

The proof in [**69**] *is by Lyapunov–Schmidt reduction arguments and the index theorem in* [**44**].

For the Bullough–Dodd equation (11.43), $\ker L = \{u_{c,x}\}$ where L is defined by (11.46). Since the momentum of the problem is $\frac{1}{2}\int (u_c')^2\, dx$, by similar computation as in (11.12), it was shown in [**69**] that

$$\langle L^{-1}u_c'', u_c'' \rangle = -\frac{1}{2}\partial_c \int (u_c')^2\, dx = -\frac{1}{2}\partial_c \left[c^{-\frac{1}{2}} \int (u_1')^2\, dx \right] > 0,$$

where $u_c = u_1(x/\sqrt{c})$ and $-u_1'' + au_1 - f(u_1) = 0$. So we get the following

THEOREM 11.4. *Assume* $f(u)$ *is a smooth function satisfying (*11.44*) and the traveling wave solution* $u_c(x - ct)$ *to (*11.43*) exists with* $c > 0$ *and* $u_c(x) \to 0$ *as* $|x| \to \infty$, *then* u_c *is linearly unstable.*

In [**69**], the above Theorem was proved for smooth and convex function f. Their additional convexity assumption on f was used to verify the condition C1) in Remark 11.6.

Besides the above linear instability result, Theorem 2.2 can be applied to give the exponential trichotomy for the linearized equation (11.45). This will be useful for the construction of invariant manifolds of (11.43) near the unstable traveling wave orbit.

11.5. Stability of steady flows of 2D Euler equation

We consider the 2D Euler equations

$$\partial_t u + (u \cdot \nabla u) + \nabla p = 0, \tag{11.49}$$

$$\nabla \cdot u = 0, \tag{11.50}$$

in a bounded domain $\Omega \subset \mathbf{R}^2$ with smooth boundary $\partial\Omega$ composed of a finite number of connected components Γ_i . The boundary condition is

$$u \cdot n = 0 \quad \text{on } \partial\Omega,$$

For simplicity, first we consider Ω to be simply connected and $\partial\Omega = \Gamma$. The vorticity form of (11.49)-(11.50) is given by

$$\partial_t \omega + \psi_y \partial_x \omega - \psi_x \partial_y \omega = 0, \tag{11.51}$$

where ψ is the stream function, then $\omega \equiv -\Delta\psi \equiv -\left(\partial_x^2 + \partial_y^2\right)\psi$ is the vorticity and $u = \nabla^\perp \psi = (\psi_y, -\psi_x)$ is the velocity. The boundary condition associated with (11.51) is given by $\psi = 0$ on $\partial\Omega$. A stationary solution of (11.51) is given by a stream function ψ_0 satisfying

$$-\psi_{0_y} \partial_x \omega_0 + \psi_{0_x} \partial_y \omega_0 = 0, \tag{11.52}$$

here $\omega_0 \equiv -\Delta\psi_0$ and $u_0 = \nabla^\perp \psi_0$ are the associated vorticity and velocity. Suppose ψ_0 satisfy the following elliptic equation

$$-\Delta\psi_0 = g(\psi_0)$$

with boundary condition $\psi_0 = 0$ on $\partial\Omega$, where g is some differentiable function. Then $\omega_0 \equiv -\Delta\psi_0 = g(\psi_0)$ is a steady solution of equation (11.51). The linearized equation near ω_0 is

$$\partial_t \omega + \psi_{0_y} \partial_x \omega - \psi_{0_x} \partial_y \omega = -\psi_y \partial_x \omega_0 + \psi_x \partial_y \omega_0, \tag{11.53}$$

with $\omega = -\Delta\psi$ and the boundary condition $\psi\,|_{\partial\Omega} = 0$. The above equation can be written as

$$\partial_t \omega + u_0 \cdot \nabla\omega - g'(\psi_0)\, u_0 \cdot \nabla\psi = 0. \tag{11.54}$$

Below we consider the case when $g' > 0$ which appeared in many interesting cases such as mean field equations (e.g. [**14**] [**15**]). Then (11.54) has the following Hamiltonian structure

$$\partial_t \omega = JL\omega, \text{ where } J = -g'(\psi_0)\, u_0 \cdot \nabla, \quad L = \frac{1}{g'(\psi_0)} - (-\Delta)^{-1}. \tag{11.55}$$

We take the energy space of the linearized Euler (11.55) as the weighted space

$$X = \{\omega \,|\, \|\omega\|_X < \infty\}, \text{ where } \|\omega\|_X = \left(\int\int_\Omega \frac{|\omega|^2}{g'(\psi_0)}\, dxdy\right)^{\frac{1}{2}}.$$

If g' has a positive lower bound, X is equivalent to $L^2(\Omega)$. In general, $\omega \in X$ implies $\omega \in L^2$ and $\nabla\psi \in L^2$. Therefore $\langle L\cdot, \cdot\rangle$ defines a bounded symmetric quadratic form on X and $L : X \to X^*$ is a bounded symmetric operator. Moreover, it is easy to see that

$$S : L^2 \to X, \quad S\omega = g'(\psi_0)^{\frac{1}{2}}\omega$$

defines an isometry. As $f(\psi_0)\cdot$ and $u_0 \cdot \nabla$ are commutative for any f, we have

$$\tilde{J} \triangleq S^{-1} J (S^*)^{-1} = -u_0 \cdot \nabla : (L^2)^* \to L^2$$

is anti-self-dual due to $\nabla \cdot u_0 = 0$, from which we obtain $J^* = -J$ and thus (**H1**) is satisfied by J and X. Moreover, since $\frac{1}{g'(\psi_0)}\cdot : X \to X^*$ is an isomorphism and

$(-\Delta)^{-1}$ is compact, we have $\dim \ker L < \infty$ and thus (**H3**) is satisfied. Note that the closed subspace $\ker J \subset X^*$ is infinite dimensional since

$$\ker J \supset \left\{h(\psi_0),\ h \in C^1\right\}.$$

Let $\tilde{P} : (L^2)^* \to \ker \tilde{J}$ be the orthogonal projection and define

$$P = (S^*)^{-1} \tilde{P} S^* : X^* \to \ker J.$$

Clearly P is a bounded linear operator on X^* and it defines a projection on X^*, but orthogonal in the L^2 sense. In fact, due to the commutativity between $f(\psi_0)\cdot$ and $u_0 \cdot \nabla$ for any f, operators P and $\tilde{P}$ take the same form shown in ([**49**])

$$P\phi\mid_{\gamma_i(c)} = \frac{\oint_{\gamma_i(c)} \frac{\phi(x,y)}{|\nabla \psi_0|} dl}{\oint_{\gamma_i(c)} \frac{1}{|\nabla \psi_0|} dl},$$

where c is in the range of ψ_0 and $\gamma_i(c)$ is a branch of $\{\psi_0 = c\}$. As in ([**49**]), define operator $A : H_0^1 \cap H^2(\Omega) \to L^2(\Omega)$ by

$$A\phi = -\Delta\phi - g'(\psi_0)\phi + g'(\psi_0) P\phi.$$

We also denote the operator

$$A_0 = -\Delta - g'(\psi_0) : H_0^1 \cap H^2(\Omega) \to L^2(\Omega)\ .$$

Clearly A, A_0 are self-adjoint with compact resolvent and thus with only discrete spectrum. The next lemma studies the spectral information of L on the weighted space X.

Recall that for any subspace $Y \in X$, $\langle L\cdot,\cdot\rangle$ also defines a bounded symmetric quadratic form on the quotient space $Y/(Y \cap \ker L)$.

LEMMA 11.3. *i) The assumption (**H2**) is satisfied by $\langle L\cdot,\cdot\rangle$ on X, with $n^-(L) = n^-(A_0)$ and* $\dim \ker L = \dim \ker A_0$.

ii) The quadratic form $\langle L\cdot,\cdot\rangle$ is non-degenerate on $\overline{R(J)}/\left(\overline{R(J)} \cap \ker L\right)$ if and only if $\ker A \subset \ker A_0$. *Moreover,*

$$n^-\left(L|_{\overline{R(J)}/\left(\overline{R(J)}\cap \ker L\right)}\right) = n^-\left(L|_{\overline{R(J)}}\right) = n^-(A). \tag{11.56}$$

PROOF. i) For any $\omega \in X$, we have

(11.57)

$$\begin{aligned}
\langle L\omega,\omega\rangle &= \int\int_\Omega \left\{\frac{\omega^2}{g'(\psi_0)} - |\nabla\psi|^2\right\} dxdy \\
&= \int\int_\Omega \left\{\frac{\omega^2}{g'(\psi_0)} - 2\psi\omega + |\nabla\psi|^2\right\} dxdy \\
&= \int\int_\Omega \left\{\left(\frac{\omega}{\sqrt{g'(\psi_0)}} - \psi\sqrt{g'(\psi_0)}\right)^2 - g'(\psi_0)\psi^2 + |\nabla\psi|^2\right\} dxdy \\
&\geq \int\int_\Omega \left[|\nabla\psi|^2 - g'(\psi_0)\psi^2\right] dxdy = (A_0\psi,\psi),
\end{aligned}$$

where $\psi = (-\Delta)^{-1}\omega$. Recall that $n^{\leq 0}(L)$ and $n^{\leq 0}(A_0)$ denote the maximal dimensions of subspaces where the quadratic forms $\langle L\cdot,\cdot\rangle$ and $(A_0\cdot,\cdot)$ are nonpositive. Let

$$\{\psi_1,\cdots,\psi_l\},\quad l = n^{\leq 0}(A_0),$$

be linearly independent eigenfunctions associated to nonpositive eigenvalues of A_0. Define the space $Y_1 \subset X$ by

$$Y_1 = \left\{ \omega \in X \mid \int_\Omega \psi_j (-\Delta)^{-1} \omega = 0,\ 1 \le j \le l \right\}.$$

Then for any $\omega \in Y_1$, we have

$$(A_0 \psi, \psi) \ge \delta \|\psi\|_{H^1}^2, \text{ for some } \delta > 0.$$

So by (11.57), for any $\omega \in Y_1$,

$$\begin{aligned} \langle L\omega, \omega \rangle &= \varepsilon \int\int_\Omega \left\{ \frac{\omega^2}{g'(\psi_0)} - |\nabla \psi|^2 \right\} dxdy + (1-\varepsilon) \langle L\omega, \omega \rangle \\ &\ge \varepsilon \int\int_\Omega \left\{ \frac{\omega^2}{g'(\psi_0)} - |\nabla \psi|^2 \right\} dxdy + (1-\varepsilon) \delta \|\psi\|_{H^1}^2 \\ &\ge \varepsilon \int\int_\Omega \left\{ \frac{\omega^2}{g'(\psi_0)} + |\nabla \psi|^2 \right\} dxdy, \end{aligned}$$

by choosing $\varepsilon > 0$ such that $(1-\varepsilon)\delta > 2\varepsilon$. Since the positive subspace Y_1 has co-dimension $n^{\le 0}(A_0)$, this shows that the assumption (**H2**) for L on X is satisfied and $n^{\le 0}(L) \le n^{\le 0}(A_0)$.

To prove $n^{\le 0}(L) \ge n^{\le 0}(A_0)$, let $\tilde{\omega}_j = g'(\psi_0)\psi_j \in X$ and $\tilde{\psi}_j = (-\Delta)^{-1} \tilde{\omega}_j$, $j = 1, \dots, l$ and then

$$\begin{aligned} (A_0 \psi_j, \psi_j) &= \int\int_\Omega \left[|\nabla \psi_j|^2 - g'(\psi_0) \psi_j^2 \right] dxdy = \int\int_\Omega \left[|\nabla \psi_j|^2 - \frac{\tilde{\omega}_j^2}{g'(\psi_0)} \right] dxdy \\ &= \int\int_\Omega \left[|\nabla \psi_j|^2 - 2\tilde{\omega}_j \psi_j + \frac{\tilde{\omega}_j^2}{g'(\psi_0)} \right] dxdy \\ &= \int\int_\Omega \left[|\nabla \psi_j|^2 - 2\nabla \psi_j \cdot \nabla \tilde{\psi}_j + \frac{\tilde{\omega}_j^2}{g'(\psi_0)} \right] dxdy \\ &\ge \int\int_\Omega \left[\frac{\tilde{\omega}_j^2}{g'(\psi_0)} - \left| \nabla \tilde{\psi}_j \right|^2 \right] dxdy = \langle L\tilde{\omega}_j, \tilde{\omega}_j \rangle, \end{aligned}$$

and thus $n^{\le 0}(L) \ge n^{\le 0}(A_0)$. Combined with above, this implies that $n^{\le 0}(L) = n^{\le 0}(A_0)$. Since $\omega \in \ker L$ if and only if $\psi = (-\Delta)^{-1} \omega \in \ker A_0$, we obtain $\dim \ker L = \dim \ker A_0$ and thus

$$n^-(L) = n^{\le 0}(L) - \dim \ker L = n^{\le 0}(A_0) - \dim \ker A_0 = n^-(A_0).$$

ii) Note that, like J, the projection P also commutes with $f(\psi_0)\cdot$ for any f. Therefore $\omega \in \overline{R(J)}$ if and only if $P\frac{\omega}{g'(\psi_0)} = 0$. It implies that

$$(I-P)L\omega = \frac{\omega}{g'(\psi_0)} - (I-P)\psi = \frac{1}{g'(\psi_0)} A\psi, \quad \forall \omega \in \overline{R(J)}, \tag{11.58}$$

where $\psi = (-\Delta)^{-1} \omega$, and thus

$$(-\Delta) \ker A = \overline{R(J)} \cap \ker \big((I-P)L \big) = \overline{R(J)} \cap \ker JL. \tag{11.59}$$

Since

$$\ker \left(\langle L\cdot, \cdot \rangle|_{\overline{R(J)}} \right) = \overline{R(J)} \cap \ker JL,$$

it immediately implies

$$\dim\ker\left(\langle L\cdot,\cdot\rangle|_{\overline{R(J)}}\right)=\dim\ker A. \tag{11.60}$$

Suppose $\langle L\cdot,\cdot\rangle$ is degenerate on $\overline{R(J)}/\left(\overline{R(J)}\cap\ker L\right)$, namely

$$\exists\,\omega_1\in\overline{R(J)}\backslash\ker L\ \text{ such that }\ \langle L\omega_1,\omega\rangle=0,\ \forall\ \omega\in\overline{R(J)}.$$

Such ω_1 satisfies $0\neq L\omega_1\in\ker J$, or equivalently $(I-P)\,L\omega_1=0$. Therefore (11.59) implies $A\psi_1=0$. Since $A_0\psi_1\neq 0$ due to $L\omega_1\neq 0$, we obtain $\ker A\subsetneqq\ker A_0$. The converse can be proved similarly and the first statement follows.

To prove (11.56), first we notice that for any $\omega\in\overline{R(J)}$,

$$\langle L\omega,\omega\rangle=\int\int_\Omega\left\{\left(\frac{\omega}{\sqrt{g'(\psi_0)}}-\psi\sqrt{g'(\psi_0)}\right)^2-g'(\psi_0)\,\psi^2+|\nabla\psi|^2\right\}dxdy$$

$$=\int\int_\Omega\Big[\left(\frac{\omega}{\sqrt{g'(\psi_0)}}-\sqrt{g'(\psi_0)}\,(I-P)\,\psi\right)^2+g'(\psi_0)\,(P\psi)^2$$
$$-g'(\psi_0)\,\psi^2+|\nabla\psi|^2\Big]\,dxdy$$

$$\geq\int\int_\Omega|\nabla\psi|^2-g'(\psi_0)\,\psi^2+g'(\psi_0)\,(P\psi)^2\,dxdy=(A\psi,\psi)\,. \tag{11.61}$$

Next, for any $\psi\in H_0^1$, let

$$\tilde{\omega}=g'(\psi_0)\,(I-P)\,\psi\in\overline{R(J)},\quad\tilde{\psi}=(-\Delta)^{-1}\tilde{\omega},$$

then

$$\begin{aligned}(A\psi,\psi)&=\int\int_\Omega|\nabla\psi|^2-g'(\psi_0)\left((I-P)\,\psi\right)^2dxdy\\&=\int\int_\Omega\left[|\nabla\psi|^2-\frac{\tilde{\omega}^2}{g'(\psi_0)}\right]dxdy\\&=\int\int_\Omega\left[|\nabla\psi|^2-2\tilde{\omega}\psi+\frac{\tilde{\omega}^2}{g'(\psi_0)}\right]dxdy\\&\geq\int\int_\Omega\left[\frac{\tilde{\omega}^2}{g'(\psi_0)}-\left|\nabla\tilde{\psi}\right|^2\right]dxdy=\langle L\tilde{\omega},\tilde{\omega}\rangle\,.\end{aligned} \tag{11.62}$$

From (11.61), (11.62) and (11.60), we get (11.56) as in the proof of i). □

By Lemma 11.3 and (iii) of Proposition 2.8, we have

THEOREM 11.5. *Assume $g'(\psi_0)>0$ and $\ker A=\{0\}$, then the index formula*

$$k_r+2k_c+2k_i^{\leq 0}=n^-(A)\,. \tag{11.63}$$

holds. In particular, when $n^-(A)$ is odd, there is linear instability; when $A>0$, there is linear stability.

PROOF. To apply (iii) of Proposition 2.8 to obtain (11.63), it suffices to verify that a.) $\langle L\cdot,\cdot\rangle$ is non-degenerate on $\ker(JL)/\ker L$, which is satisfied due to Lemma 2.1, Lemma 11.3, and $\{0\}=\ker A\subset\ker A_0$; and b.)

$$\tilde{S}\triangleq\overline{R(J)}\cap(JL)^{-1}(\ker L)=\{0\}.$$

To see the latter, we first note that $\ker A = \{0\}$ and (11.59) imply

$$\overline{R(J)} \cap \ker\big((I-P)L\big) = \{0\}.$$

Consequently, if $\omega \in \overline{R(J)} \cap (JL)^{-1}(\ker L)$, then $JL\omega \in R(J) \cap \ker L$ must vanish, namely, $L\omega \in \ker J$, and thus $(I-P)L\omega = 0$. Again, since $\omega \in \overline{R(J)}$, we obtain $\omega = 0$. Therefore $\tilde{S} = \{0\}$ and (11.63) follows.

The instability of e^{tJL} under the assumption of $n^-(A)$ being odd is straightforward from (11.63). Finally suppose $A > 0$, (11.56) and (11.60) imply that $\langle L\cdot, \cdot\rangle$ is uniformly positive definite on $\overline{R(J)}/\big(\overline{R(J)} \cap \ker L\big) = \overline{R(J)}$. Therefore e^{tJL} is stable on the closed invariant subspace $\overline{R(J)}$ and thus its stability follows from the decomposition $X = \ker(JL) + \overline{R(J)}$, which is proved in Proposition 2.8. □

By Theorem 2.6, the index formula (11.63) and the fact $i\mathbf{R} \subset \sigma(JL)$ for the linearized Euler equation imply the following.

COROLLARY 11.5. *Under the assumption of Theorem* 11.5, *when* $n^-(A) > 0$, *then there is linear instability or structural instability for* JL *(in the sense of Theorem* 2.6*).*

In the sense of Theorem 2.6, the structural instability of the linearized Euler equation $\omega_t = JL\omega$ means that there exist arbitrarily small bounded perturbations $L_{\#}$ to L such that $JL_{\#}$ has unstable eigenvalues. However, it is not clear that such perturbations can be realized in the context of the Euler equation, such as by considering neighboring steady states along with possible small domain variation.

REMARK 11.7. *In* [**49**], *it was shown that for general* $g \in C^1$, *when* $\ker A = \{0\}$ *and* $n^-(A)$ *is odd, there is linear instability. Here, the index formula (*11.63*) gives more detailed information about the spectrum of the linearized Euler operator.*

We give one example satisfying the stability condition $A > 0$. *Let* $\lambda_0 > 0$ *be the lowest eigenvalue of* $-\Delta$ *in* Ω *with Dirichlet boundary condition and* ψ_0 *be the corresponding eigenfunction. Then* $g'(\psi_0) = \lambda_0$ *and it is easy to show that* $A > 0$.

REMARK 11.8. *When the domain* Ω *is not simply connected, let* $\partial\Omega = \cup_{i=0}^n \Gamma_i$ *consist of outer boundary* Γ_0 *and* n *interior boundaries* $\Gamma_1, \cdots, \Gamma_n$. *Then the operators* A_0, A, $-\Delta$ *should be defined by using the boundary conditions:*

$$\phi|_{\Gamma_i} \text{ is constant}, \oint_{\Gamma_i} \frac{\partial \phi}{\partial n} = 0 \text{ and } \int\int_{\Omega} \phi\, dxdy = 0. \tag{11.64}$$

*The same formula (*11.63*) is still true. The linearized stream functions satisfying* (11.64) *represent perturbations preserving the circulations along each* Γ_i, *which are conserved in the nonlinear evolution.*

Below, we consider the case when $\ker A$ is nontrivial. This usually happens when the problem has some symmetry. As an example, we consider the case when Ω is a channel, that is,

$$\Omega = \{y_1 \leq y \leq y_2,\ x \text{ is } T-\text{periodic}\}.$$

The steady stream function ψ_0 satisfies

$$-\Delta\psi_0 = g(\psi_0) \text{ in } \Omega, \tag{11.65}$$

with boundary conditions ψ_0 being constants on $\{y = y_i\}$, $i = 1, 2$, where $g \in C^1$. Define the operators $L,\ A_0,\ A$ as before with the boundary conditions

$$\phi \text{ is constant on } \{y = y_i\}, \int_{\{y=y_i\}} \frac{\partial \phi}{\partial y} dx = 0,\ i = 1, 2, \tag{11.66}$$

and $\int \int_\Omega \phi \, dxdy = 0$. Taking $x-$derivative of equation (11.65), we get

$$-\Delta \psi_{0,x} = g'(\psi_0)\, \psi_{0,x} \text{ in } \Omega,$$

and $\psi_{0,x}$ satisfies the boundary condition (11.66). Thus we have $A_0 \psi_{0,x} = 0$ and

$$L\omega_{0,x} = L\left(g'(\psi_0)\, \psi_{0,x}\right) = 0.$$

Since $\psi_{0,x} = u_0 \cdot \nabla(-y)$, so $P\psi_{0,x} = 0$ and thus $A\psi_{0,x} = A_0 \psi_{0,x} = 0$.

THEOREM 11.6. *Assume $g'(\psi_0) > 0$ and $\ker A = span\{\psi_{0,x}\}$, then*

$$\exists\, \omega_1 \in \overline{R(J)} \;\; \text{such that } JL\omega_1 = \omega_{0,x}.$$

Moreover, if

$$d = \langle L\omega_1, \omega_1 \rangle = -\int \int_\Omega y\omega_1 dxdy = T(\psi_1|_{y=y_1} - \psi_1|_{y=y_2}) \neq 0,$$

where ψ_1 satisfies (11.66) *and $-\Delta \psi_1 = \omega_1$, or more explicitly*

$$\psi_1 = -A^{-1}\left(g'(\psi_0)\,(I - P)\, y\right),$$

then we have the index formula

$$k_r + 2k_c + 2k_i^{\leq 0} = n^-(A) - n^-(d), \tag{11.67}$$

where $n^-(d) = 1$ if $d < 0$ and $n^-(d) = 0$ if $d > 0$.

PROOF. Our assumption implies $\ker A \subset \ker A_0$ and thus $\langle L\cdot, \cdot \rangle$ is non-degenerate on $\overline{R(J)}/\left(\overline{R(J)} \cap \ker L\right)$ by Proposition 2.8. To apply index formula (2.20), we need to obtain the non-degeneracy of $\langle L\cdot, \cdot \rangle$ on $\tilde{S}/(\tilde{S} \cap \ker L)$ and compute $n^{\leq 0}(L|_{\tilde{S}/(\tilde{S} \cap \ker L)})$, where

$$\tilde{S} = \overline{R(J)} \cap (JL)^{-1}(\ker L) = \overline{R(J)} \cap (JL)^{-1}\left(\ker L \cap \overline{R(J)}\right).$$

Since $P\psi_{0,x} = 0$ implies $\omega_{0,x} = \frac{\psi_{0,x}}{g'(\psi_0)} \in \overline{R(J)}$. From (11.59) and our assumption on $\ker A$, we have

$$span\{\omega_{0,x}\} \subset \overline{R(J)} \cap \ker L \subset \overline{R(J)} \cap \ker\left((I - P)L\right) = \Delta \ker A = span\{\omega_{0,x}\},$$

which yields

$$\overline{R(J)} \cap \ker L = span\{\omega_{0,x}\}.$$

By the definition of $\tilde{S}$, $\omega \in \tilde{S}$ if and only if there exist $\omega \in \overline{R(J)}$ and $a \in \mathbf{R}$ such that

$$a\omega_{0,x} = JL\omega = J(I - P)L\omega.$$

Since $\omega_{0,x} = -Jy = -J(I - P)y$, we obtain equivalently $(I - P)(L\omega + ay) = 0$. From (11.58), it follows that

$$\omega \in \tilde{S} \Longleftrightarrow \omega \in \overline{R(J)} \text{ and } A\psi = -ag'(\psi_0)(I - P)y, \text{ where } -\Delta\psi = \omega.$$

Note $\ker A = span\{\psi_{0,x}\}$ and

$$\langle g'(\psi_0)\,(I - P)\, y, \psi_{0,x} \rangle = \int \int_\Omega yg'(\psi_0)\, \psi_{0,x}\, dxdy = 0.$$

There exists a stream function ψ_1 satisfying

$$A\psi_1 = -g'(\psi_0)(I-P)y,$$

which implies $\omega_1 = -\Delta\psi_1 \in \overline{R(J)}$ and $JL\omega_1 = \omega_{0,x}$. Namely $\omega_1 \in \tilde{S}$ and $\tilde{S} = span\{\omega_{0,x}, \omega_1\}$. One may compute

$$\begin{aligned} d &= \langle L\omega_1, \omega_1\rangle = \langle (I-P)L\omega_1, \omega_1\rangle = \langle -(I-P)y, \omega_1\rangle \\ &= -\langle y, \omega_1\rangle = T(\psi_1|_{y=y_1} - \psi_1|_{y=y_2}) \end{aligned}$$

where the last equal sign follows from integration by parts. If $d \neq 0$, then the desired index formula follows from (iii) of Proposition 2.8. □

Similar to Corollary 11.5 (and the comments immediately thereafter), we have

COROLLARY 11.6. *Under the assumption of Theorem 11.6, when $n^-(A) - n^-(d) > 0$, then there is linear instability or structural instability for JL.*

As another application of the Hamiltonian structure of the linearized Euler equation, we consider the inviscid damping of a stable steady flow. Assume $g'(\psi_0) > 0$ and $A > 0$, then by Theorem 11.5, the steady flow is linearly stable in the L^2 norm of vorticity. There is no time decay in $\|\omega\|_{L^2}$. However, the linear decay in the velocity norm $\|u\|_{L^2}$ is possible due to the mixing of the vorticity. For example, see [**56**] for the linear damping near Couette flow $(y,0)$ in a channel. Here, we give a weak form of the linear decay for general stable steady flows.

THEOREM 11.7. *Assume $g'(\psi_0) > 0$ and $A > 0$. For $\omega(0) \in \overline{R(J)}$, let $\omega(t) \in \overline{R(J)}$ be the solution of the linearized Euler equation (11.54). Then*

(i) When $T \to \infty$, $\frac{1}{T}\int_0^T \omega(t)\,dt \to 0$ strongly in L^2.

(ii) If there is no embedded imaginary eigenvalue of JL on $\overline{R(J)}$, then for any compact operator C in L^2, we have

$$\frac{1}{T}\int_0^T \|C\omega(t)\|_{L^2}^2\,dt \to 0, \text{ when } T \to \infty. \tag{11.68}$$

In particular, for the velocity $u = \operatorname{curl}^{-1}\omega$,

$$\frac{1}{T}\int_0^T \|u(t)\|_{L^2}^2\,dt \to 0, \text{ when } T \to \infty. \tag{11.69}$$

PROOF. By Lemma 11.3, $L|_{\overline{R(J)}} > 0$. Since $\overline{R(J)}$ is an invariant subspace of JL, we can consider the operator JL in $\overline{R(J)}$. Define the inner product $[\cdot,\cdot] = \langle L\cdot,\cdot\rangle$ on $\overline{R(J)}$, then the norm in $[\cdot,\cdot]$ is equivalent to the L^2 norm. As noted before, the operator $JL|_{\overline{R(J)}}$ is anti-self-adjoint with respect to the inner product $[\cdot,\cdot]$.

(i) By the mean ergodic convergence of unitary operators ([**73**])

$$\lim_{T\to\infty}\frac{1}{T}\int_0^T \omega(t)\,dt = \lim_{T\to\infty}\frac{1}{T}\int_0^T e^{tJL|_{\overline{R(J)}}}\omega(0)\,dt = P_0\omega(0)$$

in L^2, where P_0 is the projection operator from $\overline{R(J)}$ to $\ker JL|_{\overline{R(J)}}$ orthogonal with respect to $[\cdot,\cdot]$. Since $\ker A = \{0\}$, by Lemma 11.3 and (11.58) in particular, $\ker JL|_{\overline{R(J)}} = \{0\}$ and thus $P_0\omega(0) = 0$.

(ii) If JL has no embedded imaginary eigenvalue, then (11.68) follows directly by the RAGE theorem ([**22**]), again by using the anti-self-adjoint property of $JL|_{\overline{R(J)}}$. The conclusion (11.69) follows by choosing the compact operator $C = \mathrm{curl}^{-1}$. □

REMARK 11.9. *Assuming $A > 0$, from the proof of Theorem 11.5, the subspace $\tilde{S}$ defined in Proposition 2.8 is trivial. By Proposition 2.8, there is a direct sum decomposition $L^2 = \ker(JL) \oplus \overline{R(J)}$ invariant under JL. In fact $\ker(JL)$ corresponds to the steady solution of the linearized Euler equation. So above Lemma shows that for any initial data in L^2, in the time averaged limit, the solution of the linearized Euler equation converges to a steady solution. This is a weak form of inviscid damping.*

A stable example satisfying the assumption $A > 0$ in Theorem 11.7 is given in Remark 11.7. Below, we consider two examples of stable shear flows. First, we consider the Poisseulle flow $U(y) = y^2$ in a 2π-periodic channel $\{-1 < y < 1\}$. The linearized Euler equation becomes

$$\partial_t \omega + y^2 \partial_x \omega + 2\partial_x \psi = 0.$$

Consider the subspace of non-shear vorticities with a weighted L^2 norm

$$X_1 = \left\{ \omega = \sum_{k\in\mathbf{Z},\ k\neq 0} e^{ikx} \omega_k(y),\ \|\omega\|_{X_1}^2 = \sum_{k\in\mathbf{Z},\ k\neq 0} \|y\omega_k\|_{L^2}^2 < \infty \right\}.$$

Define $J = -\partial_x$ and $L = y^2 + 2(-\Delta)^{-1}$. Then L is uniformly positive on X_1.

Second, consider the Kolmogorov flow $U(y) = \sin y$ in a torus $T^2 = S_{\frac{2\pi}{\alpha}} \times S_{2\pi}$ with $\alpha > 1$. Here $\alpha > 1$ is the sharp stability condition since the shear flow is unstable when $\alpha < 1$. The linearized equation is

$$\partial_t \omega + \sin y \partial_x (\omega - \psi) = 0.$$

Let $J = \sin y \partial_x$ and $L = 1 - (-\Delta)^{-1}$. Then L is uniformly positive on

$$X_2 = \left\{ \omega = \sum_{k\in\mathbf{Z},\ k\neq 0} e^{ikx} \omega_k(y),\ \omega \in L^2 \right\}$$

when $\alpha > 1$. It can be shown ([**54**]) that for above two examples, the linearized Euler operator has no embedded eigenvalues. Therefore, Theorem 11.7 (ii) is true for the above two shear flows in X_1 and X_2 respectively. In particular, if we choose C to be P_N, the projection operator to the first N Fourier modes (in x), then

$$\frac{1}{T}\int_0^T \|P_N \omega(t)\|_{L^2}^2\, dt \to 0, \text{when } T \to \infty. \tag{11.70}$$

This shows that in the time averaged sense, the low frequency parts of ω tends to zero. This observation was used to prove ([**54**]) the metastability of Kolmogorov flows. In the fluid literature (see e.g. [**71**]), for 2D turbulence a dual cascade was known that energy moves to low frequency end and the enstrophy ($\int \omega^2 dx$) moves to the high frequency end. The result (11.70) can be seen as a justification of such physical intuition in a weak sense.

REMARK 11.10. *Two classes of shear flows generalizing the above two examples are studied in* [**54**]. *The linear inviscid damping in the sense of (*11.69*) is proved for stable shear flows and on the center space for the unstable shear flows, when* $\omega(0) \in L^2$ *is non-shear. Recently, for monotone and certain symmetric shear flows, more explicit linear decay estimates of the velocity were obtained in* [**74, 75, 77**] *for more regular initial data (e.g.* $\omega(0) \in H^1$ *or* H^2*).*

In [**55**]*, the stability of shear flows under Coriolis forces is studied. By using the instability index Theorem* 2.3*, the sharp stability condition for a class of shear flows can be obtained. Then the linear damping as in the above sense is proved for non-shear* $\omega(0) \in L^2$*.*

11.6. Stability of traveling waves of 2-dim nonlinear Schrödinger equations with nonzero conditions at infinity

In this section, we consider the nonlinear Schrödinger equation (NLS)

$$i\frac{\partial u}{\partial t} + \Delta u + F(|u|^2)u = 0, \quad u = u_1 + iu_2 : \mathbf{R} \times \mathbf{R}^2 \to \mathbf{C}. \tag{11.71}$$

In particular, we assume that the nonlinearity $F(s)$ satisfies

$$F \in C^2, \quad F(1) = 0, \quad F'(1) < 0. \tag{11.72}$$

Important well-known equations of this type are Gross-Pitaevskii (GP) equation with $F(s) = 1 - s$ and the cubic-quintic NLS with $F(s) = -\alpha_1 + \alpha_3 s - \alpha_5 s^2$, where α_1, α_3 and α_5 are positive constants. Assume $s = 1$ is a local minimal point of F, it is natural to consider solutions $u(t, x)$ satisfying the following boundary condition in some appropriate sense

$$|u| \to 1 \text{ as } |x| \to \infty. \tag{11.73}$$

After normalization, we can assume that $u \to 1$ when $|x| \to \infty$ in some weak sense such as $u - 1$ being approximable by Schwartz class functions in certain Sobolev norms. The equation (11.71) has the conserved energy and momentum functionals

$$E(u) = \frac{1}{2}\int_{\mathbf{R}^2} \left(|\nabla u|^2 + V(|u|^2)\right)dx,$$

$$\vec{P}(u) = (P_1(u), P_2(u)) = \frac{1}{2}\int_{\mathbf{R}^2} \langle \nabla u, i(u-1)\rangle\, dx = \int_{\mathbf{R}^2} (u_1 - 1)\nabla u_2 dx,$$

where $V(s) = \int_s^1 F(\tau)d\tau$. We also denote the first component of $\vec{P}(u)$ by

$$P(u) = \frac{1}{2}\int_{\mathbf{R}^2} \langle \partial_{x_1} u, i(u-1)\rangle\, dx = \int_{\mathbf{R}^2} (u_1 - 1)\,\partial_{x_1} u_2 dx.$$

A traveling wave (without loss of generality, in x_1-direction) of (11.71) with wave speed $c \in (0, \sqrt{2})$ is a solution in the form of $u = U_c(x_1 - ct, x_2)$, where U_c satisfies the elliptic equation

$$-ic\partial_{x_1} U_c + \Delta U_c + F(|U_c|^2)U_c = 0, \tag{11.74}$$

with the boundary condition $U_c \to 1$ when $|x| \to \infty$ in the sense $U_c - 1 \in \dot{H}^1$. Here, $\sqrt{2}$ is the sound speed and when $c \geq \sqrt{2}$, in general the traveling waves do not exist (see e.g. [**61**]). Formally, U_c is a critical point of $E - cP$. Our goal is to understand the linear stability/instability of such a traveling wave, namely, the evolution of the linearized equation of (11.71) at $U_c = u_c + iv_c$ put in the moving frame $x_1 \to x_1 - ct, x_2 \to x_2$:

$$u_t = JL_c u, \quad u = (u_1, u_2)^T \to 0 \text{ as } |x| \to \infty, \tag{11.75}$$

where $J = \begin{pmatrix} 0 & 1 \\ -1 & 0 \end{pmatrix}$ and

$$L_c := \begin{pmatrix} -\Delta - F\left(|U_c|^2\right) - 2F'\left(|U_c|^2\right) u_c^2 & -c\partial_{x_1} - 2F'\left(|U_c|^2\right) u_c v_c \\ c\partial_{x_1} - 2F'\left(|U_c|^2\right) u_c v_c & -\Delta - F\left(|U_c|^2\right) - 2F'\left(|U_c|^2\right) v_c^2 \end{pmatrix}.$$

Through L^2 duality, L_c generates the quadratic form

$$\begin{aligned} \langle L_c u, v\rangle = \int_{\mathbf{R}^2} \{\nabla u \cdot \nabla v + c(v_{1x_1} u_2 - u_1 v_{2x_1}) - F(|U_c|^2) u \cdot v \\ - 2F'(|U_c|^2)(U_c \cdot u)(U_c \cdot v)\} \, dx, \end{aligned} \tag{11.76}$$

where $u \cdot v = \mathrm{Re}(u\bar{v})$.

For the purpose of studying the linearized equation (11.75), we make the following assumptions:

(**NLS-1**) $U_c - 1 \in H^1 \times \dot{H}^1$ satisfies (11.73) and $|U_c|_{C^1(\mathbf{R}^2)} < \infty$.

(**NLS-2**) Let Γ be the collection of subspaces $S \subset H^1(\mathbf{R}^2) \times H^1(\mathbf{R}^2)$ such that $\langle L_c u, u\rangle < 0$ for all $0 \neq u \in S$, then

$$\max\{\dim S \mid S \in \Gamma\} = n^-(L_c) < \infty.$$

The above (**NLS-1**) is a natural regularity assumption. For any given traveling wave of (11.71), it is probably not so straightforward to verify (**NLS-2**). This, however, would be a direct consequence if U_c is obtained through a constrained variational approach related to energy and momentum, which is often the case. For example, in [**19**] [**20**], the 2D traveling waves of (11.74) were constructed by minimizing the functional $E(u) - cP(u)$ subject to a constraint $P(u) = p$ or $E_{kin}(u) = \int |\nabla u|^2 \, dx = k$, for general nonlinearity F. The variational problem of minimizing $E(u) - cP(u)$ subject to fixed $P(u)$ was also studied in [**7**] to construct 2D traveling waves of GP equation. Since these 2D traveling waves U_c were minimizers of $E(u) - cP(u)$ subject to one constraint, it can be shown that $n^-(L_c) \leq 1$ (see e.g. the proof of Lemma 2.7 of [**53**]). Here, we note that U_c is a critical point of $E(u) - cP(u)$ and $L_c = E''(U_c) - cP''(U_c)$.

To study the quadratic form $\langle L_c \cdot, \cdot\rangle$, obviously one may take $X = H^1(\mathbf{R}^2) \times H^1(\mathbf{R}^2)$. On the one hand, the above assumptions ensure that $L_c : X \to X^* = H^{-1} \times H^{-1}$ is bounded, satisfies $L_c^* = L_c$, and has $n^-(L_c)$ negative dimensions. On the other hand, it is easy to see that $J : X^* \to X$ is unbounded, but has a dense domain $H^1 \times H^1 \subset X^* = H^{-1} \times H^{-1}$, and satisfies $J^* = -J$. However, as the boundary condition (11.73) does not provide enough control of $|u|^2$ near $|x| = \infty$ in $\langle L_c u, u\rangle$, it is not clear that (**H2.b**) can be satisfied by any decomposition.

For (11.71) considered on $\mathbf{R}^N$, $N \geq 3$, as in [**53**], it would be possible to work on $X = H^1 \times \dot{H}^1$, where $u_1 \in H^1$ and $u_2 \in \dot{H}^1$, and verify assumptions (**H1-3**) for J and L_c based on the following two observations. Firstly, in such higher dimensions, the Gagliardo-Nirenberg inequality implies that $\dot{H}^1$ functions decay at $x = \infty$ in the L^p sense. Therefore we may reasonably strengthen the boundary condition (11.73) to $U_c \to 1$ as $|x| \to \infty$. Consequently the 'principle part' in $\langle L_c u, u\rangle$ provides the control on the $H^1 \times \dot{H}^1$ norm of u. Secondly, there are indications that U_c decays like $u_c - 1 = O(|x|^{-N})$ and $v_c = O(|x|^{1-N})$ as in the case proved for the (GP) equation in [**6**]. Along with the Hardy inequality, this allows us to control those terms in (11.76) with vanishing variable coefficients by the $H^1 \times \dot{H}^1$ norm of u. See [**53**] for more details.

The situation is much worse on $\mathbf{R}^2$ unfortunately since both of the above key observations break down on $\mathbf{R}^2$. To overcome these difficulties, our idea is to study the stability of the linearized equation (11.75) on some space roughly between $H^1 \times H^1$ and $\dot{H}^1 \times \dot{H}^1$ defined according to the properties of L_c by applying Theorem 2.7.

Let $X = H^1 \times H^1$ for (11.75) and define $Q_0, Q_1 : X \to X^*$ as

$$\langle Q_0 u, v\rangle = \operatorname{Re} \int_{\mathbf{R}^2} u\bar{v}dx, \quad \langle Q_1 u, v\rangle = \operatorname{Re} \int_{\mathbf{R}^2} (u_{x_1}\bar{v}_{x_1} + u_{x_2}\bar{v}_{x_2})\ dx,$$

namely, the L^2 and $\dot{H}^1$ duality, respectively, which satisfy (**B1**) in Section 2.6. Let $\mathbb{J} : X \to X$ be $\mathbb{J} = \begin{pmatrix} 0 & 1 \\ -1 & 0 \end{pmatrix}$. Clearly $\mathbb{J}$ satisfies (**B2**) and the unbounded operator $J = \mathbb{J}Q_0^{-1} : X^* \to X$ has the same matrix representation through the L^2 duality. As $L_c - Q_1$ consists of terms of at most one order of derivative, it satisfies (**B3**). From (**NLS-2**), there exists a subspace $S \subset X$ such that $\dim S = n^-(L_c)$ and L_c is negative definite on S. By a slight perturbation, e.g. applying the mollifier to a basis of S, we obtain a subspace $X_- \subset H^3 \times H^3$ such that $\dim X_- = n^-(L_c)$ and L_c is negative definite on X_-. Let

$$X_{\geq 0} = X_-^{\perp_{L_c}} = \{u \in X \mid \langle L_c v, u\rangle = 0\} \supset \ker L_c,$$

and

$$X_+ = \{u \in X_{\geq 0} \mid \int_{\mathbf{R}^2} u \cdot v dx = 0, \forall v \in \ker L\}.$$

Since $\dim X_- < \infty$ and L_c is negative definite on X_-, from Lemma A.2 where (**H2.b**) is not necessary (see Remark A.1), we have $X = X_- \oplus X_{\geq 0}$. It is obvious $X_{\geq 0} = X_+ \oplus \ker L_c$ and the decomposition $X = X_- \oplus \ker L_c \oplus X_+$ is L_c-orthogonal. From (**NLS-2**) and the definition of $X_\pm$, $\langle L_c u, u\rangle$ is (not necessarily uniformly) positive on X_+ and thus (**B4**) is satisfied. Finally, one may compute from the construction that

$$\ker i_{X_+}^* = Q_0(\ker L_c) \oplus L_c(X_-).$$

Since we take $X_- \subset H^3 \times H^3$ and $|U_c|_{C^1} < \infty$, (**B5**) is also satisfied. From Theorem 2.7, there exists a function space Y roughly between $X = H^1 \times H^1$ and $\dot{H}^1 \times \dot{H}^1$, an extension $L_{c,Y} : Y \to Y^*$ of L_c, and the restriction

$$J_Y : Y^* \supset D(J_Y) \to Y$$

of J, such that $(Y, L_{c,Y}, J_Y)$ satisfies assumption (**H1-3**). Therefore all our main results apply to the linearized NLS (11.75) on Y.

In the rest of this section, we assume, for some $c_0 > 0$,

(**NLS**) There exists a C^1 curve of traveling waves for c near c_0 satisfying (**NLS-1**) such that $n^-(L_{c_0}) \leq 1$ and (**NLS-2**) is satisfied for $c = c_0$.

As mentioned in the above, $n^-(L_{c_0}) \leq 1$ is satisfied if U_{c_0} is constructed as minimizers of $E - c_0 P$ subject to one constraint such as fixed $P(u)$ or $E_{kin}(u)$. We shall apply Theorem 2.3 to study the linearized equation (11.75) on Y. In order to estimate $k_0^{\leq 0}$ in the counting formula (2.13), differentiating (11.74) in x_i and we get $\ker L_{c_0} \supset \{\partial_{x_i} U_{c_0}, i = 1, 2\}$. Moreover, differentiating (11.74) in c, we have

$$L_{c_0} \partial_c U_c|_{c_0} = P'(U_{c_0}) = J^{-1} \partial_{x_1} U_{c_0},$$

and thus $J L_{c_0} \partial_c U_c|_{c_0} \in \ker L_{c_0}$. Since

$$\langle L_{c_0} \partial_c U_c|_{c_0}, \partial_c U_c|_{c_0} \rangle = \frac{dP(U_c)}{dc}|_{c_0},$$

by Proposition 2.7, we have $k_0^{\leq 0} \geq 1$ when $\frac{dP(U_c)}{dc}|_{c_0} \leq 0$ and in this case U_c is spectrally stable by (2.13).

The traveling waves constructed in the literature ([**7**] [**19**] [**20**]) are even in x_2, that is, of the form $U_c(x_1, |x_2|)$. Thus, we can consider odd and even perturbations (in x_2) respectively. We consider the even perturbations, that is, in the space $Y_e = \{u \in Y \mid u \text{ is even in } x_2\}$. For traveling waves as constrained minimizers of $E - cP$, in general it can be shown that there is at least one even negative direction of $\langle L_c \cdot, \cdot \rangle$, which then implies $n^-(L_c|_{Y_e}) = 1$. Such a symmetry preserving negative direction of L_c was constructed in [**53**] for the 3D case. For the 2D case, an even negative direction could be constructed by refining the Derrick type arguments used in [**39**]. More specifically, one can consider a scaled traveling wave $U^{a,b} = U_{c_0}(ax_1, bx_2)$ and choose a family of parameters $a(s), b(s)$ near 1 with $a(0) = b(0) = 1$ such that

$$(E - c_0 P)\left(U^{a,b}\right) < (E - c_0 P)(U_{c_0}),$$

from which an even negative direction $\frac{d}{ds} U^{a(s),b(s)}|_{s=0}$ may be obtained. If in addition to the condition $n^-(L_c|_{Y_e}) = 1$, we assume that $\partial_{x_1} U_c$ is the only even kernel of L_c, then by Theorem 2.3 and Proposition 2.7, there is linear instability in case $\frac{dP(U_c)}{dc}|_{c_0} > 0$. We summarize above discussions in the following theorem.

THEOREM 11.8. *(i) Assuming (**NLS**), the 2D traveling wave U_{c_0} is spectrally stable if $\frac{dP(U_c)}{dc}|_{c_0} \leq 0$.*

(ii) If we further assume that U_{c_0} is even in x_2 and there exists $v \in Y_e$ in the negative direction of L_{c_0} and $\ker L_{c_0} \cap Y_e = span\{\partial_{x_1} U_{c_0}\}$, then $\frac{dP(U_c)}{dc}|_{c_0} > 0$ implies linear instability of U_{c_0}.

For the GP equation, by numerical computations ([**39**]) $dP/dc < 0$ is true for the whole solitary wave branch. Thus 2D traveling waves of GP are expected to be linearly stable. In [**19**], the orbital stability of these GP traveling waves was obtained by showing concentration compactness of the constrained minimizing sequence, under the assumption of local uniqueness of minimizers. The transversal instability of 2D traveling waves of GP to 3D perturbation was proved in [**53**]. For general nonlinear term F such as cubic-quintic type, it is possible that there is an unstable branch of 2D traveling waves with $dP/dc > 0$. See the numerical examples given in [**20**].

Lastly, as a corollary of Theorems 2.3 and 11.8, we prove that the traveling waves U_{c_0} have positive momentum $P(U_{c_0})$.

COROLLARY 11.7. *Under the assumptions in both (i) and (ii) of Theorem 11.8, except for the signs of $\frac{dP(U_c)}{dc}|_{c_0}$, we have $P(U_{c_0}) > 0$.*

PROOF. First, we find v_2 such that $L_{c_0}v_2 = J^{-1}\partial_{x_2}U_{c_0}$. Consider traveling waves $U_{\vec{c}}\ (\vec{x}-\vec{c}t)$ with velocity vector $\vec{c}=(c_1,c_2)$ and $|\vec{c}| = c \in \left(0,\sqrt{2}\right)$, which satisfies

$$-J\vec{c}\cdot\nabla U_{\vec{c}} + \Delta U_{\vec{c}} + F(|U_{\vec{c}}|^2)U_{\vec{c}} = 0. \tag{11.77}$$

Let

$$Q = \frac{1}{|\vec{c}|}\begin{pmatrix} c_1 & c_2 \\ -c_2 & c_1 \end{pmatrix}$$

be the rotating matrix which transforms $\vec{c}$ to $(c,0)$, then it is easy to check that $U_{\vec{c}}(\vec{x}) = U_c(Q\vec{x})$ is a solution of (11.77) and

$$\vec{P}(U_{\vec{c}}) = Q^T\vec{P}(U_c) = P(U_c)\frac{\vec{c}}{c},$$

where we use $\vec{P}(U_c) = P(U_c)(1,0)^T$ which is due to the evenness of U_c in x_2. Differentiating (11.77) in c_2 and then evaluating at $(c_0,0)$, we get

$$L_{c_0}\partial_{c_2}U_{\vec{c}}|_{(c_0,0)} = J^{-1}\partial_{x_2}U_{c_0}.$$

Thus we can choose $v_2 = \partial_{c_2}U_{\vec{c}}|_{(c_0,0)}$ and

$$\langle L_{c_0}v_2, v_2\rangle = \partial_{c_2}P_2(U_{\vec{c}})|_{(c_0,0)} = \partial_{c_2}\left(P(U_c)\frac{c_2}{c}\right)|_{(c_0,0)} = \frac{P(U_{c_0})}{c_0}.$$

Denote $v_1 = \partial_c U_c|_{c_0}$ and recall that

$$L_{c_0}v_1 = J^{-1}\partial_{x_1}U_{c_0},\ \langle L_{c_0}v_1, v_1\rangle = \frac{dP(U_c)}{dc}|_{c_0}.$$

Also, by using the evenness of U_{c_0} in x_2, we get

$$\langle L_{c_0}v_2, v_1\rangle = \langle J^{-1}\partial_{x_2}U_{c_0}, \partial_c U_c|_{c_0}\rangle = 0,$$

and thus

$$\langle L_{c_0}\cdot,\cdot\rangle|_{span\{v_1,v_2\}} = \begin{pmatrix} \frac{dP(U_c)}{dc}|_{c_0} & 0 \\ 0 & \frac{P(U_{c_0})}{c_0} \end{pmatrix}.$$

Since

$$n^{\le 0}\left(L_{c_0}|_{span\{v_1,v_2\}}\right) \le k_0^{\le 0}(L_{c_0}) \le n^-(L_{c_0}) \le 1,$$

so when $\frac{dP(U_c)}{dc}|_{c_0} \le 0$, we must have $P(U_{c_0}) > 0$. When $\frac{dP(U_c)}{dc}|_{c_0} > 0$ and with the assumptions of Theorem 11.8 (ii), U_{c_0} is linearly unstable, which again implies that $P(U_{c_0}) > 0$. Since otherwise $P(U_{c_0}) \le 0$, then $k_0^{\le 0}(L_{c_0}) \ge 1$ and by Theorem 2.3, U_{c_0} is linearly stable, a contradiction. □

REMARK 11.11. *For 2D traveling wave solution U_c satisfying (11.74), one can prove the identity*

$$cP(U_c) = 2\int_{\mathbf{R}^2} V(|U_c|)^2\,dx, \tag{11.78}$$

by using energy conservation and virial identity (see [**20**] *for general F and* [**39**] *for GP). So for F such that V is nonnegative (such as GP), we have $P(U_c) > 0$ from (11.78). However, when V also takes negative values (such as cubic-quintic), then one can not conclude the sign of $P(U_c)$ from (11.78). By using the index counting, above Corollary 11.7 shows that $P(U_c) > 0$ is true for any nonlinear term F under the assumptions there.*

Consider axial symmetric 3D traveling waves $U_c = (x_1, |x^\perp|)$ which are constrained energy-momentum minimizers, as constructed in [**60**]. *We can also prove that $P(U_c) > 0$ by the same arguments as in Corollary* 11.7. *Actually, the argument for 3D is much simpler than 2D and does not need the additional assumptions on* $\ker L_c$. *Let*

$$v_1 = \partial_c U_c,\ v_j = \partial_{c_j} U_{\vec{c}}|_{(c,0,0)},\ j = 2, 3,$$

where $\vec{c} = (c_1, c_2, c_3)$ with $|\vec{c}| = c \in (0, \sqrt{2})$ and $U_{\vec{c}}$ is the traveling wave with the velocity vector $\vec{c}$. Then we can compute in a similar way that

$$\langle L_c \cdot, \cdot \rangle |_{span\{v_1, v_2, v_3\}} = \begin{pmatrix} \frac{dP(U_c)}{dc} & 0 & 0 \\ 0 & \frac{P(U_c)}{c} & 0 \\ 0 & 0 & \frac{P(U_c)}{c} \end{pmatrix}.$$

Since

$$n^{\leq 0} \left(L_c|_{span\{v_1, v_2, v_3\}} \right) \leq n^- (L_c) \leq 1$$

by the index counting formula (2.13)*, so regardless of the sign of $\frac{dP(U_c)}{dc}$, we must have $P(U_c) > 0$. The 3D analogue (see* [**60**]*) of the identity* (11.78) *is*

$$cP(U_c) = \int_{\mathbf{R}^3} \left| \frac{\partial U_c}{\partial x_1} \right|^2 dx + \int_{\mathbf{R}^3} V(|U_c|)^2 \, dx,$$

which is again not enough to conclude $P(U_c) > 0$ when V takes negative values.

APPENDIX A

Appendix

In this appendix, we give some elementary properties of (2.1), which are mostly based on theoretical functional analysis arguments. They include some basic decomposition of the phase space, the well-posedness of (2.1), and the standard complexification procedure.

We start with some elementary properties of L. First we prove that $n^-(L) = \dim X_-$ in assumption (**H2**) is the maximal dimension of subspaces where $\langle L\cdot,\cdot\rangle < 0$.

LEMMA A.1. *If $N \subset X$ is a subspace such that $\langle Lu, u\rangle < 0$ for all $u \in N\backslash\{0\}$, then* $\dim N \le n^-(L)$.

PROOF. Let $X_\pm$ be given in (**H2**) and $P_{+,0,-}$ be the projections associated to the decomposition $X = X_+ \oplus \ker L \oplus X_-$. For any $u \in X$, $P_- u = 0$ would imply $u \in \ker L \oplus X_+$ and thus $\langle Lu, u\rangle \ge 0$, so $u \notin N$. Therefore $P_- : N \to X_-$ is injective and in turn it implies $\dim N \le \dim X_-$. □

In order to proceed we have to introduce some notations. Given a closed subspace $Y \subset X$, let $i_Y : Y \to X$ be the embedding and then $i_Y^* : X^* \to Y^*$. Define

$$\text{(A.1)}\qquad \begin{aligned} &L_Y = i_Y^* L i_Y : Y \to Y^*, \\ &Y^{\perp_L} = \ker(i_Y^* L) = \{u \in X \mid \langle Lu, i_Y v\rangle = \langle Lu, v\rangle = 0,\ \forall v \in Y\}, \end{aligned}$$

which satisfy

$$\text{(A.2)}\qquad L_Y^* = L_Y \ \text{ and } \langle L_Y u, v\rangle = \langle Lu, v\rangle,\ \forall u, v \in Y.$$

The following is a simple technical lemma.

LEMMA A.2. *Assume **(H1-3)**. Let $Y \subset X$ be a closed subspace.*

1. *Suppose the quadratic form $\langle L\cdot,\cdot\rangle$ is non-degenerate (in the sense of* (2.4)*) on Y, then $X = Y \oplus Y^{\perp_L}$.*
2. *Assume* $\dim\ker L < \infty$ *and* $\ker L_Y = \{0\}$*, then $\langle L\cdot,\cdot\rangle$ is non-degenerate on Y.*
3. *If $X = \ker L \oplus Y$ then $\langle L\cdot,\cdot\rangle$ is non-degenerate on Y.*

PROOF. We first notice that L_Y being an isomorphism implies $Y \cap Y^{\perp_L} = \{0\}$. For any $u \in X$, let

$$u_1 = L_Y^{-1} i_Y^* L u \in Y \implies \langle Lu_1 - Lu, v\rangle = 0,\ \forall v \in Y,$$

and thus $u_2 = u - u_1 \in Y^{\perp_L}$ which implies $X = Y \oplus Y^{\perp_L}$.

In order prove the second statement, from the standard argument, it suffices to show that

$$\inf_{u\in Y\backslash\{0\}} \sup_{v\in Y\backslash\{0\}} \frac{|\langle Lu, v\rangle|}{\|u\|\|v\|} > 0. \tag{A.3}$$

According to Remark 2.2 and the assumption of the lemma, there exist closed subspaces $X_{\le 0}$ and X_+ such that the decomposition $X = X_{\le 0} \oplus X_+$ is orthogonal with respect to both $(\cdot,\cdot)$ and $\langle L\cdot,\cdot\rangle$, $\dim X_{\le 0} < \infty$, $\langle Lu, u\rangle \le 0$ for all $u \in X_{\le 0}$, and for some $\delta > 0$, $\langle Lu, u\rangle \ge \delta\|u\|^2$ for all $u \in X_+$. This splitting is associated to the orthogonal projections $\mathcal{P}_{\le 0,+} : X \to X_{\le 0,+}$. Let $Y_+ = Y \cap X_+$ and

$$Y_1 = \{u \in Y \mid \langle Lu, v\rangle = 0, \ \forall v \in Y_+\}.$$

Clearly Y_+ and Y_1 are both closed subspaces of Y. Much as in the first statement, using the uniform positive definiteness of $\langle Lu, u\rangle$ on Y_+, we have $Y = Y_+ \oplus Y_1$ via

$$u = u_+ + (u - u_+), \text{ where } u_+ = L_{Y_+}^{-1} i_{Y_+}^* Lu \in Y_+, \quad \forall u \in Y.$$

For any $u_1 \in Y_1\backslash\{0\}$, let $x_{\le 0,+} = \mathcal{P}_{X_{\le 0,+}} u_1$ and we have $u_1 = x_{\le 0} + x_+$. Since $\mathcal{P}_{X_{\le 0}} u_1 = 0$ would imply $u_1 \in X_+ \cap Y = Y_+$ contradictory to $Y = Y_+ \oplus Y_1$, we obtain that the linear mapping $\mathcal{P}_{X_{\le 0}}|_{Y_1}$ is one-to-one. Therefore $\dim Y_1 < \infty$. From the definition of Y_1, if $u_1 \in Y_1\backslash\{0\}$ satisfies that $\langle Lu_1, v\rangle = 0$ for all $v \in Y_1$, we would have $L_Y u_1 = 0$ which contradicts the assumption $\ker L_Y = \{0\}$. Therefore $L_Y|_{Y_1}$ defines an isomorphism from Y_1 to Y_1^* as $\dim Y_1 < \infty$ and thus there exists $\delta' > 0$ such that for any $u_1 \in Y_1\backslash\{0\}$, there exists $v \in Y_1$ such that $\langle L_Y u_1, v\rangle \ge \delta'\|u_1\|\|v\|$.

Consider any $u = u_1 + u_+ \in Y$. If $\|u_1\| \ge \|u_+\|$, there exists $v \in Y_1$ such that

$$\langle Lu, v\rangle = \langle Lu_1, v\rangle \ge \delta'\|u_1\|\|v\| \ge \frac{\delta'}{2}\|u\|\|v\|.$$

If $\|u_+\| \ge \|u_1\|$, then let $v = u_+$ and we have

$$\langle Lu, v\rangle = \langle Lu_+, u_+\rangle \ge \delta\|u_+\|^2 \ge \frac{\delta}{2}\|u\|\|v\|.$$

Therefore (A.3) is obtained and the second statement is proved.

Finally we prove the last statement. We first show the non-degeneracy of $\langle L\cdot,\cdot\rangle$ on $X_+ \oplus X_-$ though a standard procedure. The bounded symmetric quadratic form $\langle L\cdot,\cdot\rangle$ on $X_+ \oplus X_-$ induces bounded linear operators

$$L_{\alpha,\beta} = i_{X_\alpha}^* L i_{X_\beta} : X_\beta \to X_\alpha^*, \quad \alpha, \beta \in \{+,-\}.$$

Since L_{++} and $-L_{--}$ are both symmetric and bounded below, thus isomorphic, and $L_{+-} = L_{-+}^*$, so the same are true for

$$L_{++} - L_{+-}L_{--}^{-1}L_{-+} \quad \text{and} \quad -(L_{--} - L_{-+}L_{++}^{-1}L_{+-}).$$

It is easy to verify that

$$L^{-1} = (L_{++} - L_{+-}L_{--}^{-1}L_{-+})^{-1} i_{X_+}^* + (L_{--} - L_{-+}L_{++}^{-1}L_{+-})^{-1} i_{X_-}^*$$

is a bounded operator from $(X_+ \oplus X_-)^*$ to $X_+ \oplus X_-$. In general, if $X = \ker L \oplus Y$, there exists an isomorphism $T : X_- \oplus X_+ \to \ker L$ such that $Y = \text{graph}(T)$. The non-degeneracy of $\langle L\cdot,\cdot\rangle$ on Y follows immediately from its non-degeneracy on $X_- \oplus X_+$. The proof of the lemma is complete. □

REMARK A.1. *The first statement in the lemma holds actually for any closed subspace $Y \subset X$ as long as $\langle L\cdot, \cdot\rangle$ is non-degenerate on Y. The finite dimensionality assumption on* $\ker L$ *is essential for the second statement in the above lemma. A counter example is*

$$X = l^2 \oplus l^2, \; L = I \oplus 0, \; Y = \big\{(\{x_n\}, \{y_n\}) \in X \mid x_n = \frac{1}{n} y_n\big\},$$

for which $\dim\ker L = \infty$, $n^-(L) = 0$, $\ker L|_Y = \{0\}$, *but $\langle L\cdot, \cdot\rangle$ is not non-degenerate on Y in the sense of (2.4).*

The next lemma will allow us to decompose equation (2.1).

LEMMA A.3. *Suppose $X_{1,2} \subset X$ are closed subspaces satisfying $X = X_1 \oplus X_2$. Let $P_{1,2} : X \to X_{1,2}$ be the associated projections, which imply $P_{1,2}^* : X_{1,2}^* \to X^*$, and*

$$J_{jk} = P_j J P_k^* : D(J_{jk}) \to X_j, \; D(J_{jk}) = (P_k^*)^{-1}\big(D(J) \cap P_k^* X_k^*\big), \; j, k = 1, 2.$$

1. *If* $\ker i_{X_2}^* \subset D(J)$, *then J_{11} and J_{21} are bounded operators defined on X_1^*, $J_{11}^* = -J_{11}$, $J_{22} = -J_{22}^*$, and $J_{12}^* = -J_{21}$, and J_{12} can be extended to the bounded operator $-J_{21}^* = J_{12}^{**}$ defined on X_2^*.*
2. *If $\langle Lu_1, u_2\rangle = 0$, for all $u_j \in X_j$, $j = 1, 2$, then $LX_j \subset$* $\ker i_{X_{3-j}}^*$, *$L_{X_{1,2}}$ satisfy **(H2)** on $X_{1,2}$, $n^-(L) = n^-(L_{X_1}) + n^-(L_{X_2})$, and* $\ker L = \ker L_{X_1} \oplus \ker L_{X_2}$.
3. *Assume $\langle Lu_1, u_2\rangle = 0$, for all $u_j \in X_j$, $j = 1, 2$, and* $\ker i_{X_2}^* \subset D(J)$, *then the combinations (X_j, L_{X_j}, J_{jj}), $j = 1, 2$, satisfy **(H1-3)**.*

PROOF. For $j = 1, 2$, define $\tilde{X}_j^*$ as

$$\tilde{X}_j^* = P_j^* X_j^* = \ker i_{X_{3-j}}^* = \{f \in X^* \mid \langle f, u\rangle = 0, \; \forall u \in X_{3-j}\} \subset X^*. \tag{A.4}$$

Clearly it holds

$$i_{X_1} P_1 + i_{X_2} P_2 = I_X, \quad P_1^* i_{X_1}^* + P_2^* i_{X_2}^* = I_{X^*}, \quad X^* = \tilde{X}_1^* \oplus \tilde{X}_2^*. \tag{A.5}$$

Assume $\tilde{X}_1^* = P_1^* X_1^* \subset D(J)$. The Closed Graph Theorem implies that the closed operator $JP_1^* : X_1^* \to X$ is actually bounded, and thus J_{11} and J_{21} are bounded as well. The property $J_{11}^* = -J_{11}$ is obvious from $J^* = -J$ and the boundedness of J_{11}. We also obtain from this assumption and (A.5) that $D(J) \cap \tilde{X}_2^*$ is dense in $\tilde{X}_2^*$ and thus J_{12} and J_{22} are densely defined, as $P_j^* : X_j^* \to \tilde{X}_j^*$ is an isomorphism. It remains to prove $J_{12}^* = -J_{21}$ and $J_{22}^* = -J_{22}$.

Suppose $u = J_{12}^* g$, or equivalently, $g \in X_1^*$ and $u \in X_2$ satisfy, $\forall f \in D(J_{12}) \subset X_2^*$,

$$\langle P_2^* f, i_{X_2} u - i_{X_1} P_1 J P_1^* g\rangle = \langle f, u\rangle = \langle g, J_{12} f\rangle = \langle P_1^* g, J P_2^* f\rangle, \tag{A.6}$$

where we used $P_j i_{X_j} = id$ and $P_{3-j} i_{X_j} = 0$ on X_j. For any $h \in X_1^*$, we have

$$\langle P_1^* h, i_{x_2} u - i_{X_1} P_1 J P_1^* g\rangle = \langle P_1^* g, J P_1^* h\rangle.$$

Therefore, (A.5) and (A.6) imply $u = J_{12}^* g$ is equivalent to

$$\begin{aligned} &\langle \gamma, i_{X_2} u - i_{X_1} P_1 J P_1^* g\rangle = \langle P_1^* g, J\gamma\rangle, \quad \forall \gamma \in D(J) \\ \Longleftrightarrow\; & i_{X_2} u - i_{X_1} P_1 J P_1^* g = J^* P_1^* g = -J P_1^* g \\ \Longleftrightarrow\; & u = -P_2 J P_1^* g = -J_{21} g \end{aligned}$$

Therefore $J_{12}^* = -J_{21}$.

Similarly, using the assumption $\tilde{X}_1^* \subset D(J)$, one can prove $u = J_{22}^* g \in X_2$, $g \in X_2^*$, if and only if

$$i_{X_2} u + i_{X_1} J_{21}^* g = J^* P_2^* g \Longleftrightarrow u = -P_2 J P_2^* g = -J_{22} g.$$

Therefore we obtain $J_{22}^* = -J_{22}$.

Assume $\langle Lu_1, u_2 \rangle = 0$, for all $u_j \in X_j$, $j = 1, 2$. As a direct consequence, we have $LX_j \subset \tilde{X}_j^*$, which, along with (A.5), immediately implies

$$L = P_1^* L_{X_1} P_1 + P_2^* L_{X_2} P_2, \quad P_j^* L_{X_j} P_j(X) \subset \tilde{X}_j,$$

which in turn yield

$$\ker L = \ker L_{X_1} \oplus \ker L_{X_2}, \quad \ker L_{X_j} = X_j \cap \ker L, \ j = 1, 2.$$

Let

$$Y_{1,2} = \{u \in X_{1,2} \mid (u, v) = 0, \ \forall v \in \ker L_{X_{1,2}}\}, \ Y = Y_1 \oplus Y_2,$$

which implies

$$X = Y \oplus \ker L = Y_1 \oplus Y_2 \oplus \ker L,$$

and

$$\langle Ly_j, y_1' + y_2' + u \rangle = \langle Ly_j, y_j' \rangle = \langle L_{Y_j} y_j, y_j' \rangle,$$

for any $y_j, y_j' \in Y_j$, $j = 1, 2$, and $u \in \ker L$. Let $P_{Y_{1,2,0}}$ be the projections associated to this decomposition, then we have

$$L(Y_j) \subset \tilde{Y}_j^* \triangleq P_{Y_j}^* Y_j^* = \ker i_{\ker L \oplus Y_{3-j}}^*, \quad j = 1, 2.$$

Assumption (**H2**) implies that $L|_Y : Y \to R(L)$ is an isomorphism to the closed subspace $R(L) \subset X^*$. Therefore, $L(Y_{1,2}) \subset \tilde{Y}_{1,2}^*$ are closed subspaces and $L|_{Y_{1,2}} : Y_{1,2} \to L(Y_{1,2})$ are isomorphisms. It implies that $L_{Y_{1,2}}$ are isomorphisms from $Y_{1,2}$ to closed subspaces $L_{Y_{1,2}}(Y_{1,2}) \subset Y_{1,2}^*$. Due to their boundedness and symmetry, we obtain that $L_{Y_{1,2}} Y_{1,2}$ is equal to the orthogonal complement of $\ker L_{Y_{1,2}}^* = \ker L_{Y_{1,2}} = \{0\}$. So $L_{Y_{1,2}} : Y_{1,2} \to Y_{1,2}^*$ are isomorphisms, which induce bounded non-degenerate symmetric quadratic forms on $Y_{1,2}$. From the standard theory on symmetric quadratic forms, Y_j, $j = 1, 2$, can be split into $Y_j = Y_{j+} \oplus Y_{j-}$, where closed subspaces $Y_{j\pm}$ are orthogonal with respect to both $(\cdot, \cdot)$ and $\langle L\cdot, \cdot \rangle$. Moreover, there exists $\delta > 0$ such that

$$\pm \langle L_{X_j} u, u \rangle = \pm \langle Lu, u \rangle \geq \delta \|u\|^2, \ \forall u \in Y_{j\pm}.$$

This proves that X_j satisfies (**H2**) with

$$X_j = Y_{J-} \oplus \ker L_{X_j} \oplus Y_{j+}, \quad j = 1, 2.$$

Finally, since $X = X_1 \oplus X_2$, there exists $C > 0$ such that,

$$\|u_1\|^2 + \|u_2\|^2 \leq C \|u_1 + u_2\|^2, \quad \forall \ u_{1,2} \in X_{1,2}.$$

Therefore the splitting

$$X = (Y_{1-} \oplus Y_{2-}) \oplus \ker L \oplus (Y_{1+} \oplus Y_{2+})$$

satisfies the properties in (**H2**), which implies $n^-(L) = n^-(L_{X_1}) + n^-(L_{X_2})$.

Finally, assume $\langle Lu_1, u_2 \rangle = 0$, for all $u_j \in X_j$, $j = 1, 2$, and $P_1^* X_1^* \subset D(J)$. To complete the proof of the lemma, we only need to show that (**H3**) is satisfied by (X_j, L_{X_j}, J_{jj}), $j = 1, 2$. This is obvious for $j = 1$, as J_{11} is a bounded operator, and thus we only need to work on $j = 2$. Let $X_\pm \subset X$ be the closed subspaces assumed in (**H2-3**) and $Z = X_- \oplus X_+$. Since $X = \ker L \oplus Z = \ker L \oplus Y$, Z can be represented as the graph of a bounded linear operator from Y to $\ker L$.

As $\ker L = \ker L_{X_1} \oplus \ker L_{X_2}$ and $Y = Y_1 \oplus Y_2$, there exist bounded operators $S_{jk} : Y_k \to \ker L_{X_j}$ such that

$$Z = \{y_1 + y_2 + \Sigma_{j,k=1}^{2} S_{jk} y_k \mid y_{1,2} \in Y_{1,2}\}.$$

We will first show

$$W \triangleq \{f \in X_2^* \mid \langle f, u\rangle = 0,\, u \in Z_2\} \subset D(J_{22}), \tag{A.7}$$

where

$$Z_2 = \{y_2 + S_{22} y_2 \mid y_2 \in Y_2\} \subset X_2.$$

Trivially extend S_{jk} to be an operator from X_k to $\ker L_{X_j} \subset X_j$ via

$$S_{jk}(y_k + v_k) = S_{jk} y_k, \quad \forall\, y_k \in Y_k,\ v_k \in \ker L_{X_k}.$$

It leads to $S_{jk} S_{kl} = 0$, $\forall j, k, l = 1, 2$. Given any $f \in W \subset X_2^*$, one may compute, for any

$$u = y_1 + y_2 + \Sigma_{j,k=1}^{2} S_{jk} y_k \in Z,$$

using the definition of W, and the property of the extensions of S_{jk},

$$\begin{aligned}\langle P_2^* f - P_1^* S_{21}^* f, u\rangle =&\langle f, y_2 + S_{21} y_1 + S_{22} y_2\rangle - \langle S_{21}^* f, y_1 + S_{11} y_1 + S_{12} y_2\rangle \\ =&\langle f, S_{21} y_1\rangle - \langle f, S_{21} y_1 + S_{21} S_{11} y_1 + S_{21} S_{12} y_2\rangle = 0.\end{aligned}$$

Therefore **(H3)** implies $P_2^* f - P_1^* S_{21}^* f \in D(J)$. Since we assume $P_1^* X_1^* \subset D(J)$, we obtain $P_2^* f \in D(J)$ and thus $f \in D(J_{22})$ which proves (A.7).

Since $y_2 \to y_2 + S_{22} y_2$ is an isomorphism from Y_2 to Z_2,

$$\langle L(y_2 + S_{22} y_2), y_2' + S_{22} y_2'\rangle = \langle L y_2, y_2'\rangle,$$

and L_{Y_2} is isomorphic, we have $\langle L\cdot, \cdot\rangle$ is non-degenerate on Z_2 and L_{Z_2} is also an isomorphism. Therefore there exist closed subspaces $X_{2\pm} \subset Z_2$ and $\delta > 0$ such that $Z_2 = X_{2-} \oplus X_{2+}$, $\dim X_{2-} = n^-(L_{X_2})$, and $\pm\langle L_{X_2} u, u\rangle \geq \delta \|u\|^2$, for any $u \in X_{2\pm}$. It along with (A.7) and $X_2 = Z_2 \oplus \ker L_{X_2}$ completes the proof of the lemma. □

REMARK A.2. *Under assumptions $\langle L u_1, u_2\rangle = 0$, for all $u_j \in X_j$, $j = 1, 2$, and $P_1^* X_1^* \subset D(J)$, (X_j, L_{X_j}, J_{jj}), $j = 1, 2$, satisfies the same hypothesis **(H1-H3)** as (X, L, J) and $n^-(L) = n^-(L_{X_1}) + n^-(L_{X_2})$. Moreover, it is easily verified based on these assumptions that $J_{jj} L_{X_j} = P_j J L|_{X_j}$. Therefore this lemma would often be applied to reduce the problem to subspaces when $JL(X_1) \subset X_1$, which implies JL has certain upper triangular structure.*

COROLLARY A.1. *$LJ : D(J) \to X^*$ is a closed operator and consequently $(JL)^* = -LJ$.*

PROOF. Let $X_\pm$ and $\ker L$ satisfy the requirements in **(H2-3)** and let $X_1 = \ker L$ and $X_2 = X_- \oplus X_+$. Clearly we have, $L_{X_1} = 0$, $\langle L u_1, u_2\rangle = 0$, for all $u_j \in X_j$, $j = 1, 2$, and $P_1^* X_1^* \subset D(J)$ due to **(H3)**. Using

$$i_{X_1} P_1 + i_{X_2} P_2 = I_X, \quad L i_{X_1} = 0, \quad i_{X_1}^* L = 0,$$

LJ can be rewritten in this decomposition

$$LJ\gamma = P_2^* L_{X_2} J_{21} i_{X_1}^* \gamma + P_2^* L_{X_2} J_{22} i_{X_2}^* \gamma, \quad \forall \gamma \in X^*,$$

which is equivalent to using the blockwise decomposition of J and L. Since J_{21} is continuous, $P_2^* L_{X_2} J_{21} i_{X_1}^*$ is continuous too. Moreover, the facts that $L_{X_2} : X_2 \to X_2^*$ is an isomorphism, P_2^* has a continuous left inverse $i_{X_2}^*$ as $P_2 i_{X_2} = I_{X_2}$, along with the closedness of J_{22} imply that $P_2^* L_{X_2} J_{22}$ and thus $P_2^* L_{X_2} J_{22} i_{X_2}^*$ is a closed operator. Therefore LJ is closed.

Since $(LJ)^* = JL$ is densely defined and thus $(LJ)^{**} = -(JL)^*$ is well defined. The closeness of LJ implies $LJ = (LJ)^{**} = -(JL)^*$. □

REMARK A.3. *We would like to point out that, in the proof Lemma* A.3 *and Corollary* A.1, *we do not use the assumption that $n^-(L) < \infty$. Therefore, they actually hold even if $n^-(L) = \infty$ except that $n^-(L_{X_{1,2}})$ might be ∞.*

The following is a simple, but useful, technical lemma.

LEMMA A.4. *There exist closed subspaces $X_\pm \subset X$ satisfying the properties in (**H2-3**) and in addition,*

(1) *$X = X_0 \oplus X_- \oplus X_+$ is a L-orthogonal splitting with associated projections $P_{0,\pm}$, where $X_0 = \ker L$;*
(2) *$L_{X_\pm} : X_\pm \to X_\pm^*$ are isomorphic; and*
(3) *$\tilde{X}_{0,-}^* \subset D(J)$ and $D(J) \cap \tilde{X}_+^*$ is dense in $\tilde{X}_+^*$, where $\tilde{X}_{\pm,0}^* \triangleq P_{\pm,0}^* X_{\pm,0}^*$ (see* (A.1) *and* (A.4)*).*

PROOF. Let $Y_\pm \subset X$ be closed subspaces satisfying hypothesis (**H2-3**). Let $Y = Y_- \oplus Y_+$, $P : X \to Y$ be the projection associated to the decomposition $X = X_0 \oplus Y$, $\tilde{X}_0^* = (I - P)^* X_0^*$, and $\tilde{Y}^* = P^* Y^*$, which are closed subspaces. According to (**H3**), we have $\tilde{X}_0^* \subset D(J)$. Consequently, $\tilde{Y}^* \cap D(J)$ is dense in $\tilde{Y}^*$ as $X^* = \tilde{X}_0^* \oplus \tilde{Y}^*$. Our assumptions imply $L_Y : Y \to \tilde{Y}^*$ is an isomorphism, which induces a bounded symmetric quadratic form on Y with Morse index equal to $n^-(L)$. Therefore there exists a closed subspace $X_- \subset Y$ such that $\dim X_- = n^-(L)$, $L(X_-) \subset D(J)$, and $\langle Lu, u\rangle \leq -\delta\|u\|^2$, for all $u \in X_-$. Let

$$X_+ = \{u \in Y \mid \langle Lu, v\rangle = 0, \forall v \in X_-\}.$$

Since L is uniformly negative on X_-, Lemma A.2 implies the L-orthogonal splitting $Y = X_- \oplus X_+$ and thus the L-orthogonal decomposition $X = X_0 \oplus X_- \oplus X_+$ as well. The rest of the proof follows easily from the facts that L_Y is isomorphic, $X^* = \tilde{X}_0^* \oplus \tilde{X}_-^* \oplus \tilde{X}_+^*$, $\dim X_- = n^-(L)$, $\tilde{X}_0^* \subset D(J)$, and $\tilde{X}_-^* = L(X_-) \subset D(J)$. □

REMARK A.4. *Under assumption* (2.2), *it is possible to choose $X_\pm$ such that $X_+ \oplus X_- = (\ker L)^\perp$ satisfies all properties in Lemma* A.4, *where $(\ker L)^\perp$ is defined in* (2.3). *In fact, let $Y = (\ker L)^\perp$, then* (2.2) *implies that the splitting $X = \ker L \oplus Y$ satisfies all assumptions in Lemma* A.3. *The rest of the construction of $X_\pm \subset Y = (\ker L)^\perp$ follows in exactly the same procedure as in the proof of Lemma* A.4.

In order to establish the well-posedness of the linear equation in the next, we start with the following lemma.

LEMMA A.5. *There exists an equivalent inner product $(\cdot,\cdot)_L$ on X, a linear operator $A : D(JL) \to X$ which is anti-self-adjoint with respect to $(\cdot,\cdot)_L$, and a bound linear operator $B : X \to X$ such that $JL = A + B$.*

PROOF. Let $X = X_- \oplus X_0 \oplus X_+$ be a decomposition as given in Lemma A.4 with $X_0 = \ker L$. Let

$$L_\pm = \pm P_\pm^* i_{X_\pm}^* L i_{X_\pm} P_\pm : X \to X^*,$$

which satisfy

$$L_\pm^* = L_\pm, \quad L = L_+ - L_-, \quad \langle L_\pm u, v\rangle = \pm\langle L_{X_\pm} u, v\rangle = \pm\langle Lu, v\rangle, \, \forall u, v \in X_\pm.$$

Let $R : X \to X^*$ be the isomorphism corresponding to $(\cdot, \cdot)$ through the Riesz Representation Theorem and

$$L_0 = P_0^* i_{X_0}^* R i_{X_0} P_{X_0} : X \to X^* \quad \longleftrightarrow \quad \langle L_0 u, v\rangle = (P_0 u, P_0 v).$$

From Lemma A.4 and assumptions (**H2-3**), it is easy to verify that

$$(u, v)_L \triangleq \langle (L_+ + L_- + L_0)u, v\rangle = \langle L_+ u, v\rangle + \langle L_- u, v\rangle + (P_0 u, P_0 v)$$

is uniformly positive and defines an equivalent inner product on X. Let

$$A = J(L_+ + L_- + L_0) = JL + 2JL_- + JL_0 \triangleq JL - B.$$

Since $P_{0,-}^* X_{0,-}^* \subset D(J)$, the Closed Graph Theorem implies that B is bounded. If $\dim \ker L < \infty$, B is obviously of finite rank. The proof of the lemma is complete. □

A direct consequence of this lemma is the well-posedness of equation (2.1) which follows from the standard perturbation theory of semigroups.

PROPOSITION A.1. *JL generates a C^0 group e^{tJL} of bounded linear operators on X.*

Complexification. For considerations where complex eigenvalues are involved, we have to work with the standard complexification of X and the associated operators. Let

$$\tilde X = \{x = x_1 + ix_2 \mid x_{1,2} \in X\} \text{ with } \overline{x_1 + ix_2} = x_1 - ix_2$$

equipped with the complexified inner product

$$(x_1 + ix_2, x_1' + ix_2') = (x_1, x_1') + (x_2, x_2') + i\big((x_2, x_1') - (x_1, x_2')\big).$$

Instead of complexifying L as a linear operator directly, it is much more convenient for us to complexify its corresponding real symmetric quadratic form $\langle Lu, v\rangle$ into a complex Hermitian symmetric form

$$\begin{aligned} \mathcal{B}(x_1' + ix_2', x_1 + ix_2) &= \langle \tilde L(x_1 + ix_2), (x_1' + ix_2')\rangle \\ &= \langle Lx_1, x_1'\rangle + \langle Lx_2, x_2'\rangle + i\big(\langle Lx_1, x_2'\rangle - \langle Lx_2, x_1'\rangle\big), \end{aligned} \tag{A.8}$$

for any $x_{1,2}, \, x_{1,2}' \in X$. Accordingly L is complexified to a (anti-linear) mapping $\tilde L$ from $\tilde X$ to $\tilde X^*$ satisfying

$$\tilde L(cx + c'x') = \bar c \tilde L x + \bar c' \tilde L x'. \tag{A.9}$$

A similar complexification can also be carried out for J corresponding to a Hermitian symmetric form on $\tilde X^*$ and a (anti-linear) mapping from $\tilde X^* \to \tilde X$.

The composition $\tilde J \circ \tilde L$ of (anti-linear) mappings $\tilde J$ and $\tilde L$ is a closed complex linear operator from $D(\tilde J \tilde L) \subset \tilde X$ to $\tilde X$. The fact that $\widetilde{JL}$ is anti-symmetric with respect to the Hermitian symmetric form $\langle \tilde L u, v\rangle$, that is,

$$\langle \tilde L(\tilde J \tilde L u), v\rangle = -\langle \tilde L u, \tilde J \tilde L v\rangle, \tag{A.10}$$

will be used frequently. According to Corollary A.1, the dual operator of $\tilde J \tilde L$ is given by

$$(\tilde J \tilde L)^* = -\tilde L \tilde J. \tag{A.11}$$

It is easy to verify that $\tilde{L}$, $\tilde{J}$, $\tilde{J}\tilde{L}$ and $\tilde{L}\tilde{J}$ are real in the sense

$$\overline{\langle \tilde{L}x, x' \rangle} = \langle \tilde{L}\bar{x}, \overline{x'} \rangle, \ \overline{\langle f, \tilde{J}g \rangle} = \langle \bar{f}, \tilde{J}\bar{g} \rangle, \ \overline{\tilde{J}\tilde{L}x} = \tilde{J}\tilde{L}\bar{x}, \ \overline{\tilde{L}\tilde{J}x} = \tilde{L}\tilde{J}\bar{x}. \tag{A.12}$$

This implies that the spectrum of $\tilde{J}\tilde{L}$ and $\tilde{L}\tilde{J}$ are symmetric about the real axis in the complex plane.

REMARK A.5. *In fact, on the complexified Hilbert space* $\tilde{X}$ *(or on* $\tilde{X}^*$*), a linear operator or a Hermitian form is the complexification of a (real) operator or a symmetric quadratic form on* X *(or on* $\tilde{X}^*$*) if and only if* (A.12) *holds.*

In the rest of the paper, with slight abuse of notations, we will write X, JL, $\langle Lu, v \rangle$ also for their complexifications unless confusion might occur.

REMARK A.6. *The linear group of bounded operators* e^{tJL} *obtained in Proposition* A.1 *is also complexified accordingly when needed.*

REMARK A.7. *Exactly the same statements in Lemma* A.2, 3.1, 3.2 *hold in the complexified framework.*

Bibliography

[1] J. C. Alexander and R. Sachs, *Linear instability of solitary waves of a Boussinesq-type equation: a computer assisted computation*, Nonlinear World **2** (1995), no. 4, 471–507. MR1360872

[2] J. Angulo Pava, *Nonlinear dispersive equations*, Mathematical Surveys and Monographs, vol. 156, American Mathematical Society, Providence, RI, 2009. Existence and stability of solitary and periodic travelling wave solutions, DOI 10.1090/surv/156. MR2567568

[3] J. Angulo Pava, J. L. Bona, and M. Scialom, *Stability of cnoidal waves*, Adv. Differential Equations **11** (2006), no. 12, 1321–1374. MR2276856

[4] P. W. Bates and C. K. R. T. Jones, *Invariant manifolds for semilinear partial differential equations*, Dynamics reported, Vol. 2, Dynam. Report. Ser. Dynam. Systems Appl., vol. 2, Wiley, Chichester, 1989, pp. 1–38. MR1000974

[5] T. B. Benjamin, *The stability of solitary waves*, Proc. Roy. Soc. (London) Ser. A **328** (1972), 153–183, DOI 10.1098/rspa.1972.0074. MR0338584

[6] F. Béthuel, P. Gravejat, and J.-C. Saut, *Existence and properties of travelling waves for the Gross-Pitaevskii equation*, Stationary and time dependent Gross-Pitaevskii equations, Contemp. Math., vol. 473, Amer. Math. Soc., Providence, RI, 2008, pp. 55–103, DOI 10.1090/conm/473/09224. MR2522014

[7] F. Béthuel, P. Gravejat, and J.-C. Saut, *Travelling waves for the Gross-Pitaevskii equation. II*, Comm. Math. Phys. **285** (2009), no. 2, 567–651, DOI 10.1007/s00220-008-0614-2. MR2461988

[8] S. Benzoni-Gavage, C. Mietka, and L. M. Rodrigues, *Co-periodic stability of periodic waves in some Hamiltonian PDEs*, Nonlinearity **29** (2016), no. 11, 3241–3308, DOI 10.1088/0951-7715/29/11/3241. MR3567090

[9] J. L. Bona, P. E. Souganidis, and W. A. Strauss, *Stability and instability of solitary waves of Korteweg-de Vries type*, Proc. Roy. Soc. London Ser. A **411** (1987), no. 1841, 395–412. MR897729

[10] J. L. Bona and R. L. Sachs, *Global existence of smooth solutions and stability of solitary waves for a generalized Boussinesq equation*, Comm. Math. Phys. **118** (1988), no. 1, 15–29. MR954673

[11] J. Bronski, M. A. Johnson, and T. Kapitula, *An instability index theory for quadratic pencils and applications*, Comm. Math. Phys. **327** (2014), no. 2, 521–550, DOI 10.1007/s00220-014-1949-5. MR3183408

[12] J. C. Bronski, V. M. Hur, and M. A. Johnson, *Modulational instability in equations of KdV type*, New approaches to nonlinear waves, Lecture Notes in Phys., vol. 908, Springer, Cham, 2016, pp. 83–133, DOI 10.1007/978-3-319-20690-5_4. MR3408757

[13] J. C. Bronski, M. A. Johnson, and T. Kapitula, *An index theorem for the stability of periodic travelling waves of Korteweg-de Vries type*, Proc. Roy. Soc. Edinburgh Sect. A **141** (2011), no. 6, 1141–1173, DOI 10.1017/S0308210510001216. MR2855892

[14] E. Caglioti, P.-L. Lions, C. Marchioro, and M. Pulvirenti, *A special class of stationary flows for two-dimensional Euler equations: a statistical mechanics description*, Comm. Math. Phys. **143** (1992), no. 3, 501–525. MR1145596

[15] E. Caglioti, P.-L. Lions, C. Marchioro, and M. Pulvirenti, *A special class of stationary flows for two-dimensional Euler equations: a statistical mechanics description. II*, Comm. Math. Phys. **174** (1995), no. 2, 229–260. MR1362165

[16] S.-N. Chow, X.-B. Lin, and K. Lu, *Smooth invariant foliations in infinite-dimensional spaces*, J. Differential Equations **94** (1991), no. 2, 266–291, DOI 10.1016/0022-0396(91)90093-O. MR1137616

[17] S.-N. Chow and K. Lu, *Invariant manifolds for flows in Banach spaces*, J. Differential Equations **74** (1988), no. 2, 285–317, DOI 10.1016/0022-0396(88)90007-1. MR952900

[18] M. Chugunova and D. Pelinovsky, *Count of eigenvalues in the generalized eigenvalue problem*, J. Math. Phys. **51** (2010), no. 5, 052901, 19, DOI 10.1063/1.3406252. MR2666968

[19] D. Chiron and M. Mariş, *Traveling waves for nonlinear Schrödinger equations with nonzero conditions at infinity*, Arch. Ration. Mech. Anal. **226** (2017), no. 1, 143–242, DOI 10.1007/s00205-017-1131-2. MR3686002

[20] D. Chiron and C. Scheid, *Travelling waves for the nonlinear Schrödinger equation with general nonlinearity in dimension two*, J. Nonlinear Sci. **26** (2016), no. 1, 171–231, DOI 10.1007/s00332-015-9273-6. MR3441277

[21] S. Cuccagna, D. Pelinovsky, and V. Vougalter, *Spectra of positive and negative energies in the linearized NLS problem*, Comm. Pure Appl. Math. **58** (2005), no. 1, 1–29, DOI 10.1002/cpa.20050. MR2094265

[22] H. L. Cycon, R. G. Froese, W. Kirsch, and B. Simon, *Schrödinger operators with application to quantum mechanics and global geometry*, Springer Study Edition, Texts and Monographs in Physics, Springer-Verlag, Berlin, 1987. MR883643

[23] K. F. Gurski, R. Kollár, and R. L. Pego, *Slow damping of internal waves in a stably stratified fluid*, Proc. R. Soc. Lond. Ser. A Math. Phys. Eng. Sci. **460** (2004), no. 2044, 977–994, DOI 10.1098/rspa.2003.1226. MR2133852

[24] I. Ekeland, *Convexity methods in Hamiltonian mechanics*, Ergebnisse der Mathematik und ihrer Grenzgebiete (3) [Results in Mathematics and Related Areas (3)], vol. 19, Springer-Verlag, Berlin, 1990, DOI 10.1007/978-3-642-74331-3. MR1051888

[25] K. Fan, *Invariant subspaces of certain linear operators*, Bull. Amer. Math. Soc. **69** (1963), 773–777, DOI 10.1090/S0002-9904-1963-11028-9. MR0158268

[26] F. Gesztesy, C. K. R. T. Jones, Y. Latushkin, and M. Stanislavova, *A spectral mapping theorem and invariant manifolds for nonlinear Schrödinger equations*, Indiana Univ. Math. J. **49** (2000), no. 1, 221–243, DOI 10.1512/iumj.2000.49.1838. MR1777032

[27] E. Grenier, *On the nonlinear instability of Euler and Prandtl equations*, Comm. Pure Appl. Math. **53** (2000), no. 9, 1067–1091, DOI 10.1002/1097-0312(200009)53:9⟨1067::AID-CPA1⟩3.3.CO;2-H. MR1761409

[28] M. Grillakis, *Analysis of the linearization around a critical point of an infinite-dimensional Hamiltonian system*, Comm. Pure Appl. Math. **43** (1990), no. 3, 299–333, DOI 10.1002/cpa.3160430302. MR1040143

[29] M. Grillakis, J. Shatah, and W. Strauss, *Stability theory of solitary waves in the presence of symmetry. I*, J. Funct. Anal. **74** (1987), no. 1, 160–197, DOI 10.1016/0022-1236(87)90044-9. MR901236

[30] M. Grillakis, J. Shatah, and W. Strauss, *Stability theory of solitary waves in the presence of symmetry. II*, J. Funct. Anal. **94** (1990), no. 2, 308–348, DOI 10.1016/0022-1236(90)90016-E. MR1081647

[31] Y. Guo and W. A. Strauss, *Instability of periodic BGK equilibria*, Comm. Pure Appl. Math. **48** (1995), no. 8, 861–894, DOI 10.1002/cpa.3160480803. MR1361017

[32] M. Hărăguş and T. Kapitula, *On the spectra of periodic waves for infinite-dimensional Hamiltonian systems*, Phys. D **237** (2008), no. 20, 2649–2671, DOI 10.1016/j.physd.2008.03.050. MR2514124

[33] E. Hille and R. S. Phillips, *Functional analysis and semi-groups*, American Mathematical Society Colloquium Publications, vol. 31, American Mathematical Society, Providence, R. I., 1957. rev. ed. MR0089373

[34] V. M. Hur and M. A. Johnson, *Stability of periodic traveling waves for nonlinear dispersive equations*, SIAM J. Math. Anal. **47** (2015), no. 5, 3528–3554, DOI 10.1137/12090215X. MR3397429

[35] I. S. Iohvidov, *On the spectra of Hermitian and unitary operators in a space with indefinite metric* (Russian), Doklady Akad. Nauk SSSR (N.S.) **71** (1950), 225–228. MR0035923

[36] J. Jin, S. Liao, and Z. Lin, *Nonlinear modulational instability of dispersive PDE models*, Arch. Ration. Mech. Anal. **231** (2019), no. 3, 1487–1530, DOI 10.1007/s00205-018-1303-8. MR3902468

[37] J. Jin, Z. Lin, and C. Zeng, *Dynamics near the solitary waves of the supercritical gKDV equations*, J. Differential Equations **267** (2019), no. 12, 7213–7262, DOI 10.1016/j.jde.2019.07.019. MR4011044

[38] J. Jin, Z. Lin, and C. Zeng, *Invariant manifolds of traveling waves of the 3D Gross-Pitaevskii equation in the energy space*, Comm. Math. Phys. **364** (2018), no. 3, 981–1039, DOI 10.1007/s00220-018-3189-6. MR3875821

[39] C. A. Jones, S. J. Putterman, and P. Roberts, H., *Motions in a Bose condensate V. Stability of solitary wave solutions of nonlinear Schrödinger equations in two and three dimensions.* J. Phys. A, Math. Gen., **19** (1986), 2991–3011.

[40] C. A. Jones, and P. H. Roberts, *Motions in a Bose condensate IV. Axisymmetric solitary waves.* J. Phys. A, Math. Gen., **15** (1982), 2599–2619.

[41] M. A. Johnson, *Stability of small periodic waves in fractional KdV-type equations*, SIAM J. Math. Anal. **45** (2013), no. 5, 3168–3193, DOI 10.1137/120894397. MR3116644

[42] T. Kato, *Perturbation theory for linear operators*, Classics in Mathematics, Springer-Verlag, Berlin, 1995. Reprint of the 1980 edition. MR1335452

[43] T. Kapitula and B. Deconinck, *On the spectral and orbital stability of spatially periodic stationary solutions of generalized Korteweg–de Vries equations*, Hamiltonian partial differential equations and applications, Fields Inst. Commun., vol. 75, Fields Inst. Res. Math. Sci., Toronto, ON, 2015, pp. 285–322, DOI 10.1007/978-1-4939-2950-4_10. MR3445506

[44] T. Kapitula, P. G. Kevrekidis, and B. Sandstede, *Counting eigenvalues via the Krein signature in infinite-dimensional Hamiltonian systems*, Phys. D **195** (2004), no. 3-4, 263–282, DOI 10.1016/j.physd.2004.03.018. MR2089513

[45] T. Kapitula, P. G. Kevrekidis, and B. Sandstede, *Addendum: "Counting eigenvalues via the Krein signature in infinite-dimensional Hamiltonian systems" [Phys. D* ***195*** *(2004), no. 3-4, 263–282; MR2089513]*, Phys. D **201** (2005), no. 1-2, 199–201, DOI 10.1016/j.physd.2004.11.015. MR2118638

[46] T. Kapitula and A. Stefanov, *A Hamiltonian-Krein (instability) index theory for solitary waves to KdV-like eigenvalue problems*, Stud. Appl. Math. **132** (2014), no. 3, 183–211, DOI 10.1111/sapm.12031. MR3180620

[47] R. Kollár and P. D. Miller, *Graphical Krein signature theory and Evans-Krein functions*, SIAM Rev. **56** (2014), no. 1, 73–123, DOI 10.1137/120891423. MR3246298

[48] M. G. Kreĭn, *A new application of the fixed-point principle in the theory of operators in a space with indefinite metric* (Russian), Dokl. Akad. Nauk SSSR **154** (1964), 1023–1026. MR0169059

[49] Z. Lin, *Some stability and instability criteria for ideal plane flows*, Comm. Math. Phys. **246** (2004), no. 1, 87–112, DOI 10.1007/s00220-004-1045-3. MR2044892

[50] Z. Lin, *Nonlinear instability of ideal plane flows*, Int. Math. Res. Not. **41** (2004), 2147–2178, DOI 10.1155/S107379280414018X. MR2078852

[51] Z. Lin, *Nonlinear instability of periodic BGK waves for Vlasov-Poisson system*, Comm. Pure Appl. Math. **58** (2005), no. 4, 505–528, DOI 10.1002/cpa.20028. MR2119867

[52] Z. Lin, *Instability of nonlinear dispersive solitary waves*, J. Funct. Anal. **255** (2008), no. 5, 1191–1224, DOI 10.1016/j.jfa.2008.06.003. MR2455496

[53] Z. Lin, Z. Wang, and C. Zeng, *Stability of traveling waves of nonlinear Schrödinger equation with nonzero condition at infinity*, Arch. Ration. Mech. Anal. **222** (2016), no. 1, 143–212, DOI 10.1007/s00205-016-0998-7. MR3519968

[54] Z. Lin and M. Xu, *Metastability of Kolmogorov flows and inviscid damping of shear flows*, Arch. Ration. Mech. Anal. **231** (2019), no. 3, 1811–1852, DOI 10.1007/s00205-018-1311-8. MR3902476

[55] Z. Lin, J. Yang, and H. Zhu, *Barotropic instability of shear flows*, Stud. Appl. Math. **144** (2020), no. 3, 289–326, DOI 10.1111/sapm.12297. MR4075921

[56] Z. Lin and C. Zeng, *Inviscid dynamical structures near Couette flow*, Arch. Ration. Mech. Anal. **200** (2011), no. 3, 1075–1097, DOI 10.1007/s00205-010-0384-9. MR2796139

[57] Z. Lin and C. Zeng, *Unstable manifolds of Euler equations*, Comm. Pure Appl. Math. **66** (2013), no. 11, 1803–1836, DOI 10.1002/cpa.21457. MR3103911

[58] Y. Liu, *Instability of solitary waves for generalized Boussinesq equations*, J. Dynam. Differential Equations **5** (1993), no. 3, 537–558, DOI 10.1007/BF01053535. MR1235042

[59] R. S. MacKay, *Stability of equilibria of Hamiltonian systems*, Nonlinear phenomena and chaos (Malvern, 1985), Malvern Phys. Ser., Hilger, Bristol, 1986, pp. 254–270. MR854706

[60] M. Mariş, *Traveling waves for nonlinear Schrödinger equations with nonzero conditions at infinity*, Ann. of Math. (2) **178** (2013), no. 1, 107–182, DOI 10.4007/annals.2013.178.1.2. MR3043579

[61] M. Mariş, *Nonexistence of supersonic traveling waves for nonlinear Schrödinger equations with nonzero conditions at infinity*, SIAM J. Math. Anal. **40** (2008), no. 3, 1076–1103, DOI 10.1137/070711189. MR2452881

[62] K. Nakanishi and W. Schlag, *Invariant manifolds and dispersive Hamiltonian evolution equations*, Zurich Lectures in Advanced Mathematics, European Mathematical Society (EMS), Zürich, 2011, DOI 10.4171/095. MR2847755

[63] R. L. Pego and M. I. Weinstein, *Eigenvalues, and instabilities of solitary waves*, Philos. Trans. Roy. Soc. London Ser. A **340** (1992), no. 1656, 47–94, DOI 10.1098/rsta.1992.0055. MR1177566

[64] L. Pontrjagin, *Hermitian operators in spaces with indefinite metric* (Russian, with English summary), Bull. Acad. Sci. URSS. Sér. Math. [Izvestia Akad. Nauk SSSR] **8** (1944), 243–280. MR0012200

[65] D. E. Pelinovsky, *Spectral stability on nonlinear waves in KdV-type evolution equations*, Nonlinear physical systems, Mech. Eng. Solid Mech. Ser., Wiley, Hoboken, NJ, 2014, pp. 377–400. MR3330810

[66] M. Renardy, *On the linear stability of hyperbolic PDEs and viscoelastic flows*, Z. Angew. Math. Phys. **45** (1994), no. 6, 854–865, DOI 10.1007/BF00952081. MR1306936

[67] Y. Shizuta, *On the classical solutions of the Boltzmann equation*, Comm. Pure Appl. Math. **36** (1983), no. 6, 705–754, DOI 10.1002/cpa.3160360602. MR720591

[68] M. Stanislavova and A. Stefanov, *Linear stability analysis for travelling waves of second order in time PDE's*, Nonlinearity **25** (2012), no. 9, 2625–2654, DOI 10.1088/0951-7715/25/9/2625. MR2967119

[69] M. Stanislavova and A. Stefanov, *On the spectral problem $\mathcal{L}u = \lambda u'$ and applications*, Comm. Math. Phys. **343** (2016), no. 2, 361–391, DOI 10.1007/s00220-015-2542-2. MR3477342

[70] P. E. Souganidis and W. A. Strauss, *Instability of a class of dispersive solitary waves*, Proc. Roy. Soc. Edinburgh Sect. A **114** (1990), no. 3-4, 195–212, DOI 10.1017/S0308210500024380. MR1055544

[71] P. Tabeling, *Two-dimensional turbulence: a physicist approach*, Phys. Rep. **362** (2002), no. 1, 1–62, DOI 10.1016/S0370-1573(01)00064-3. MR1898594

[72] I. Vidav, *Spectra of perturbed semigroups with applications to transport theory.*, J. Math. Anal. Appl. **30** (1970), 264–279, DOI 10.1016/0022-247X(70)90160-5. MR0259662

[73] K. Yosida, *Functional analysis*, Classics in Mathematics, Springer-Verlag, Berlin, 1995. Reprint of the sixth (1980) edition, DOI 10.1007/978-3-642-61859-8. MR1336382

[74] D. Wei, Z. Zhang, and W. Zhao, *Linear inviscid damping for a class of monotone shear flow in Sobolev spaces*, Comm. Pure Appl. Math. **71** (2018), no. 4, 617–687, DOI 10.1002/cpa.21672. MR3772399

[75] D. Wei, Z. Zhang, and W. Zhao, *Linear inviscid damping and vorticity depletion for shear flows*, Ann. PDE **5** (2019), no. 1, Art. 3, 101, DOI 10.1007/s40818-019-0060-9. MR3919496

[76] E. Zeidler, *Nonlinear functional analysis and its applications. I*, Springer-Verlag, New York, 1986. Fixed-point theorems; Translated from the German by Peter R. Wadsack, DOI 10.1007/978-1-4612-4838-5. MR816732

[77] C. Zillinger, *Linear inviscid damping for monotone shear flows in a finite periodic channel, boundary effects, blow-up and critical Sobolev regularity*, Arch. Ration. Mech. Anal. **221** (2016), no. 3, 1449–1509, DOI 10.1007/s00205-016-0991-1. MR3509006

Editorial Information

To be published in the *Memoirs*, a paper must be correct, new, nontrivial, and significant. Further, it must be well written and of interest to a substantial number of mathematicians. Piecemeal results, such as an inconclusive step toward an unproved major theorem or a minor variation on a known result, are in general not acceptable for publication.

Papers appearing in *Memoirs* are generally at least 80 and not more than 200 published pages in length. Papers less than 80 or more than 200 published pages require the approval of the Managing Editor of the Transactions/Memoirs Editorial Board. Published pages are the same size as those generated in the style files provided for $\mathcal{A}_{\mathcal{M}}\mathcal{S}$-LaTeX or $\mathcal{A}_{\mathcal{M}}\mathcal{S}$-TeX.

Information on the backlog for this journal can be found on the AMS website starting from `http://www.ams.org/memo`.

A Consent to Publish is required before we can begin processing your paper. After a paper is accepted for publication, the Providence office will send a Consent to Publish and Copyright Agreement to all authors of the paper. By submitting a paper to the *Memoirs*, authors certify that the results have not been submitted to nor are they under consideration for publication by another journal, conference proceedings, or similar publication.

Information for Authors

Memoirs is an author-prepared publication. Once formatted for print and on-line publication, articles will be published as is with the addition of AMS-prepared frontmatter and backmatter. Articles are not copyedited; however, confirmation copy will be sent to the authors.

Initial submission. The AMS uses Centralized Manuscript Processing for initial submissions. Authors should submit a PDF file using the Initial Manuscript Submission form found at `www.ams.org/submission/memo`, or send one copy of the manuscript to the following address: Centralized Manuscript Processing, MEMOIRS OF THE AMS, 201 Charles Street, Providence, RI 02904-2294 USA. If a paper copy is being forwarded to the AMS, indicate that it is for *Memoirs* and include the name of the corresponding author, contact information such as email address or mailing address, and the name of an appropriate Editor to review the paper (see the list of Editors below).

The paper must contain a *descriptive title* and an *abstract* that summarizes the article in language suitable for workers in the general field (algebra, analysis, etc.). The *descriptive title* should be short, but informative; useless or vague phrases such as "some remarks about" or "concerning" should be avoided. The *abstract* should be at least one complete sentence, and at most 300 words. Included with the footnotes to the paper should be the 2010 *Mathematics Subject Classification* representing the primary and secondary subjects of the article. The classifications are accessible from `www.ams.org/msc/`. The Mathematics Subject Classification footnote may be followed by a list of *key words and phrases* describing the subject matter of the article and taken from it. Journal abbreviations used in bibliographies are listed in the latest *Mathematical Reviews* annual index. The series abbreviations are also accessible from `www.ams.org/msnhtml/serials.pdf`. To help in preparing and verifying references, the AMS offers MR Lookup, a Reference Tool for Linking, at `www.ams.org/mrlookup/`.

Electronically prepared manuscripts. The AMS encourages electronically prepared manuscripts, with a strong preference for $\mathcal{A}_{\mathcal{M}}\mathcal{S}$-LaTeX. To this end, the Society has prepared $\mathcal{A}_{\mathcal{M}}\mathcal{S}$-LaTeX author packages for each AMS publication. Author packages include instructions for preparing electronic manuscripts, samples, and a style file that generates the particular design specifications of that publication series. Though $\mathcal{A}_{\mathcal{M}}\mathcal{S}$-LaTeX is the highly preferred format of TeX, author packages are also available in $\mathcal{A}_{\mathcal{M}}\mathcal{S}$-TeX.

Authors may retrieve an author package for *Memoirs of the AMS* from `www.ams.org/journals/memo/memoauthorpac.html`. The *AMS Author Handbook* is available in PDF format from the author package link. The author package can also be obtained free of charge by sending email to tech-support@ams.org or from the Publication Division, American Mathematical Society, 201 Charles St., Providence, RI 02904-2294, USA. When requesting an author package, please specify $\mathcal{A}\mathcal{M}\mathcal{S}$-LaTeX or $\mathcal{A}\mathcal{M}\mathcal{S}$-TeX and the publication in which your paper will appear. Please be sure to include your complete mailing address.

After acceptance. The source files for the final version of the electronic manuscript should be sent to the Providence office immediately after the paper has been accepted for publication. The author should also submit a PDF of the final version of the paper to the editor, who will forward a copy to the Providence office.

Accepted electronically prepared files can be submitted via the web at www.ams.org/submit-book-journal/, sent via FTP, or sent on CD to the Electronic Prepress Department, American Mathematical Society, 201 Charles Street, Providence, RI 02904-2294 USA. TeX source files and graphic files can be transferred over the Internet by FTP to the Internet node ftp.ams.org (130.44.1.100). When sending a manuscript electronically via CD, please be sure to include a message indicating that the paper is for the *Memoirs*.

Electronic graphics. Comprehensive instructions on preparing graphics are available at www.ams.org/authors/journals.html. A few of the major requirements are given here.

Submit files for graphics as EPS (Encapsulated PostScript) files. This includes graphics originated via a graphics application as well as scanned photographs or other computer-generated images. If this is not possible, TIFF files are acceptable as long as they can be opened in Adobe Photoshop or Illustrator.

Authors using graphics packages for the creation of electronic art should also avoid the use of any lines thinner than 0.5 points in width. Many graphics packages allow the user to specify a "hairline" for a very thin line. Hairlines often look acceptable when proofed on a typical laser printer. However, when produced on a high-resolution laser imagesetter, hairlines become nearly invisible and will be lost entirely in the final printing process.

Screens should be set to values between 15% and 85%. Screens which fall outside of this range are too light or too dark to print correctly. Variations of screens within a graphic should be no less than 10%.

Any graphics created in color will be rendered in grayscale for the printed version unless color printing is authorized by the Managing Editor and the Publisher. In general, color graphics will appear in color in the online version.

Inquiries. Any inquiries concerning a paper that has been accepted for publication should be sent to memo-query@ams.org or directly to the Electronic Prepress Department, American Mathematical Society, 201 Charles St., Providence, RI 02904-2294 USA.

Editors

This journal is designed particularly for long research papers, normally at least 80 pages in length, and groups of cognate papers in pure and applied mathematics. Papers intended for publication in the *Memoirs* should be addressed to one of the following editors. The AMS uses Centralized Manuscript Processing for initial submissions to AMS journals. Authors should follow instructions listed on the Initial Submission page found at www.ams.org/memo/memosubmit.html.